Designing Electronic Product Enclosures

Tony Serksnis

Designing Electronic Product Enclosures

 Springer

Tony Serksnis
1305 Longfellow Way
San Jose, California, USA

ISBN 978-3-030-09887-2   ISBN 978-3-319-69395-8   (eBook)
https://doi.org/10.1007/978-3-319-69395-8

This Springer imprint is published by the registered company Springer International Publishing AG part
of Springer Nature.
The registered company address is: Gewerbestrasse 11, 6330 Cham, Switzerland

Acknowledgments

I want to thank the following people and organizations for their help with this text:

- The entire team at Springer Publishing. First of all, for them believing in my message and helping me to bring it forward. The content reviewers have made many valuable contributions – they made the book better! The proofing group made my words make much greater sense.
- The Technical Illustrator for this text, Jack Warner of Meep design LLC. I worked with Jack at Trimble, where he was the first full-time Industrial Designer for Trimble. His excellence and flair are remarkable.
- All the management team and Mechanical Engineers I worked with at Trimble in my 25 years there. Charles Trimble interviewed me before I started, and I still like my answer to him on how I would save money on a project—and that would be to "design parts so that they could be tooled effectively." Trimble gave me much opportunity to design parts, and later attain, train, and manage new MEs. Many of these MEs have become Silicon Valley's best designers at Apple, Fitbit, Intuitive Surgical, and other top corporations.
- Finally, to my best (ever) girlfriend, Cindy Jeung. She has always encouraged me to do my best and to make sure I am having fun.

Pictoglyphs

Included in the text are a series of 13 "Pictoglyphs." I'm choosing the term "pictoglyph" to describe the 3 inch × 3 inch paintings (acrylic on canvas) that I painted to add some color to the book.

I wanted to have one "image" that would "boil-down" a complete chapter. This would be an image that either means something to me personally, or at least possibly leave the reader with a visual cue.

The following are some words that (attempt to) explain the pictoglyph that is with each chapter: (listed by Chapter Title, Image 3-Letter Title, and what I was trying to picture)

Chapter 1: Successful Design

PRD

What is shown is a Surveyor's Pole, GPS Receiver, and Data Recorder, using the yellow and blue tones of Trimble Inc. A Product Requirement Document (PRD) is the written statement of the product which will provide the designer a contract to proceed with the design and to provide a measurement as to how well the design has "succeeded."

Chapter 2: Building the Design

POB

The beginning point of a design, the Point of Beginning (POB, surveyor term), is shown. Two "random objects," a red and a blue "brick" are given to the designer. Right away, the designer is given the choice of how to arrange just those two objects. How far apart will they be and how large will the resulting enclosure be? Thus, the design begins and there will likely be many more "bricks" with all sorts of impending constraints.

Chapter 3: Structural Considerations

MOI

MOI stands for Moment of Inertia. The MOI of a cross-sectional area is a measure of that cross-section's ability to withstand a load. I have shown a cross-section of an I-Beam in "rusted-red" hues. An I-Beam is a standard structural member that possesses an extremely high MOI for its relatively low weight. Thus, it stands as a great example of what a designer is trying to optimize: high strength/low weight/low cost. The design of an electronic enclosure must certainly pass the structural considerations that it will be exposed to.

Chapter 4: Materials and Processes

MAT

Any design of an enclosure will involve materials. Those materials will need to be processed and formed using the most cost-effective fabrication methodologies. Thus, MAT stands for "material." The Pictograph shows the "cool steel blue" of an object being mass-produced. I tried to show a manufacturing process that repeats the shape of an object. The best choice of a material will be a balance between cost, the time allotted to bring the product to the market, and how well the product conforms to its design intent.

Chapter 5: User Interface

EYE

The human eye is a magnificent "user interface" between the brain and the outside world. The enclosure designer must allow various interfaces between the product and the user to allow that user to utilize the product. I have chosen to show a single key from a keyboard membrane switch in cross-section. The key itself is a "translucent clear material" with a red icon on top of it and a printed circuit board (green laminate) below. The key involves the human touch and visual senses to interact with the product.

Chapter 6: Assembly and Service

DFA

Design For Assembly. An enclosure designer will be designing a product that has many parts. How easily these parts go-together will determine assembly costs. The pictoglyph shows an initial assembly idea (in boring grey) consisting of two plates, spacers and hardware. This has been simplified (shown in "exciting" purple) with parts that will simplify the assembly cost and reduce the overall cost. Design for Serviceability would be a natural extension of considering costs of the product design for the life of the design.

Chapter 7: Product Environments

ENV

One of my first pictoglyphs, and it is one of my favorites, is this one. All products go into an environment, and the designer must understand that environment for the product to survive. The pictoglyph shows a surveyor's pole dropping from an upright position to the ground. We were designing carbon graphite poles to replace aluminum poles. Many questions loomed about the potential material change. These included: strength, fabrication, cost, straightness, and the interconnects needed. We had to attach sensors and get drop test information to access our design. Another positive aspect is that we invited students from Cal Poly SLO to participate in the testing as a Trimble-funded student project.

The glyph shows a yellow product on the top of a pole with a red accelerometer attached midway. A blue sky contains some of the equations involving mass and force.

Chapter 8: Cooling Techniques

QΔT

Most products need energy to do work. This energy generally produces heat that must be dissipated from the product. The designer can use either passive or forced cooling techniques, and must understand the cooling techniques available to them. Q (quantity of heat) is a function of some factor times the temperature difference (the "delta T"). The pictoglyph shows a velocity profile for laminar flow turning into turbulent flow. Can you feel the heat?… can you see the Nusselt number… ?

Chapter 9: EMC

EMC

Electromagnetic compatibility (EMC) is a general problem for designing products that need shielding against EMI/RFI. The electronics in a product must be shielded from outside sources, while itself generating limited EMI/RFI to agency approval standards. What is shown in the pictoglyph is the joint in a yellow sheet metal enclosure. A "wave disturbance" (shown in textured "invisible") is smoothed-out on the inside of the enclosure. This illustrates a design that shields against EMI/RFI.

Chapter 10: Safety by Design

The "Image 3-letter title" for this chapter is actually the four-color NFPA Hazard Diamond. This is used to identify gas, fire, and chemical information to the public.

The pictoglyph itself is rather abstract. I came across a picture of five people conferring at a construction site. Two people have orange safety vests, while three people have green safety vests. They all have white hardhats on. As seen from above, the construction site has piles of metal strewn about. This is an example of people talking about safety before proceeding with the job at hand.

Chapter 11: Shipping and Packaging

RSC

A <u>R</u>egular <u>S</u>lotted <u>C</u>ontainer is an off-the-shelf solution for a shipping box. A well-designed custom shipping container can provide the customer with protection from the shipping environment and also provide an outstanding out-of-box experience for the customer. My pictoglyph shows a brown package being machine drop-tested. The lab walls are cool mint and blue. Also shown is a standard cardboard box cross-section showing its optimal strength-to-weight ratio.

Chapter 12: Documentation

ECN

This "glyph" is an abstraction of a portion of a drawing that utilizes the Geometric Tolerancing principles of ASME Y14.5. Various "features" are shown in the primary colors of red, yellow, and blue. All product designers will produce documentation to provide information on their designs to the other people that will help bring these designs to the customer. This documentation will have changes (corrections or improvements) that are made in a controlled and orderly manner, with some system that involves an <u>E</u>ngineering <u>C</u>hange <u>N</u>otice.

Chapter 13: Continuous Improvement

DNA

What I have depicted in this pictoglyph is a chart showing "some measure" of how well a project proceeds, vs. ongoing time that a company is in business. The "DNA" of the project processes should allow a feedback system to improve the process. The brown tones turn to gold with the best practices. The product designer, besides working on the immediate electronic enclosure, should also be spending their time on how to improve the *process* of designing *all* electronic enclosures.

1305 Longfellow Way, San Jose, California, USA   Tony Serksnis

Presented here is a pictoglyph that shows all of the 3-letter titles of the book chapters:

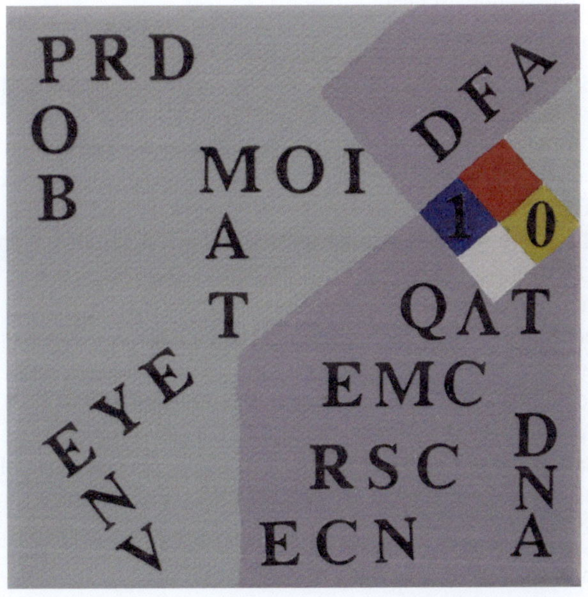

Introduction: Intent of Text

This text has been written for the developer of a product idea who needs to turn that idea into a physical piece of hardware. This developer has basically invented either some software or an electrical circuit that will reside on a printed circuit board. That circuit board will need an enclosure for a customer to be able to utilize this new idea.

Does this electrical product need an enclosure? No, not always. The Raspberry Pi Single Board Computer is an example of such a product. Originally the uncertified version sold in small numbers as a development tool. Developers would take the "RPi" and come up with commercial devices that *needed* to be both FCC and CE certified. As thousands of the RPi were sold, it was clear that the best overall solution was to get the RPi itself to pass FCC/CE (which it did in 2012). However, the use (of most) of this text is still germane, even to that product, as an RPi Single Board Computer would still need:

- Successful design – defining the design team and success
- Building the design – the printed circuit board layout and optimal object placement
- Structural considerations – mounting hole placement
- Materials and processes – materials and finishes needed for the PCB and components (PCBA)
- User interfaces – how will customer power and get signal data from PCBA
- Assembly and serviceability – optimization needed for the PCBA
- Product environments – standards and testing required
- Cooling techniques – heat analysis needed of PCBA
- EMC – standards and testing required to pass FCC/CE
- Safety by design – safety review needed in any design project
- Shipping and packaging – needed to ship PCBA to customer
- Documentation – required for purchasing components and assembling PCBA
- Continuous improvement – always needed in any design project to enhance the next project

Thus, even in the case of the "lack of enclosure," the text is useful in the design of electronic products. The above 13 bullet points make up the text outline. The main thrust of the text will be on the "commercialization" of most electronic products, that is, when an enclosure is needed on the product to meet the environmental and certification requirements of the marketplace.

A prototype enclosure will need to be produced to further that idea for that product. At some point in the product development process, the enclosure will be analyzed for its ability to:

1. Pass the structural and environmental considerations that the product will be exposed to
2. Pass the regulations needed to legally sell that product into a marketplace

The developer of the new product idea either has the basic skill set to produce that enclosure or will be contracting out for that service. If they do contract out for the design of the enclosure, the more that the developer understands about the concerns that will exist with a design (the structural, environmental, and regulatory issues), the faster the design will be able to successfully enter the marketplace. This text is intended to highlight the basic considerations of any enclosure design that contains a printed circuit board, hence the title *Designing Electronic Product Enclosures*.

For example, if the product is intended to be a new cell phone, it's clear that the sooner the enclosure and its contents will survive a drop from a person's pocket to the floor, will have a display screen that shows up in temperatures between some normal operating temperatures (-20 to $+120$ °F), will pass local and regional product safety regulations, and will pass FCC regulations for allowable electromagnetic interference (EMI) to other electronic equipment, the sooner the product can be successful in the marketplace.

This text has also been written for the mechanical engineer or designer tasked to design an enclosure for an electronic product. This text takes that person thru the considerations that must be made to produce that enclosure from the early prototype stage to the production of high quantities of product.

It is not the intent of the text to provide a complete understanding of some of the disciplines mentioned (strength of materials, shock and vibration, heat transfer), but rather to provide a reference on how these key disciplines may be considered in the design of most electronic enclosures.

Chapter Organization

Each chapter will start with a short paragraph briefly providing an introduction and explaining the logical progression of sections. This paragraph basically explains the path that is being taken towards the end goal of a successful design of an electronic product enclosure.

Each chapter will end with a short paragraph briefly summarizing its salient thoughts. This summary will be the main "take-aways" of the chapter.

The book is presented in a "building-block" form, that is, we'll start with some definitions of success for the design and then start adding some basics about layout, structural considerations, and material/process options available to the designer. With that foundation, we'll present the user interfaces that can be expected to be utilized. We'll move on to the design's consideration of assembly and serviceability. Chapter 7 will provide the basics of designing for the various environments that the products must survive in. We'll move to adding the building blocks of cooling and add the two more considerations of EMC and safety for our designs. Towards the end of the text, we'll discuss two "less glamorous" but necessary aspects of our design (shipping and documentation), before ending on a chapter on the need for design continuous improvement.

The ordering of Chaps. 3, 4, 5, 6, 7, 8, 9 and 10 is a bit arbitrary. That is, Chap. 10 "Safety" being presented *after* Chap. 3 "Structural Considerations" is not meant to imply that the subject of safety considerations is any less important. Chapters 3,

4, 5, 6, 7, 8, 9 and 10 are actually considered "all at the same time" in the EPE designer's head, and it is only after the consideration of *all* of the issues in these chapters can a design be considered as truly thought out.

References and Information Access

Each chapter will be followed by a list of references which I have found invaluable in my work. In the course of my career, there have been changes in the way reference material is accessed. In 1980, the designer used their reference texts, various company (internal and external) catalogs/design guides, and many notes kept in handwritten files and binders. This has been replaced by the ability to have most of this reference material accessible via the Internet or the organization's intranet.

A designer should have at his/her disposal a rather complete set of references. For example, one of the best sources of information on the cooling of electronic enclosures is the Pamotor Company's Catalog F117-0583. This catalog has a 26-page fan selection section that is simply packed with solid, practical information. (This catalog was acquired from a fan distributor at no charge.) I usually choose to look this catalog over before proceeding with a cooling design. A recent Internet search for "fan cooling for computer" revealed very good general information on fan selection criteria at www.nmbtc.com.

Knowledge and understanding are acquired in an incredible number of ways. Experience is, of course, the number one teacher. Our peers, or people who have "traveled down the same road" as we, are also invaluable sources of knowledge. Formal education is also essential to get beyond the "feeling" of a sound design. A deep understanding of the principles involved provides a foundation for research into new answers required for the challenges ahead in designing electronic product enclosures.

Author's Credentials

By listing some of the corporations where I have worked, this should provide some background on where my experience comes from.

Lincoln Electric Company I started in the drafting room running blue prints in 1964. By taking night school classes in drafting, I soon was given the chance to apprentice as a tool and die designer where I got experience in the basic skills of designing jigs and fixtures. Lincoln Electric is "world famous" for their profit sharing program and work ethic. (I would later teach beginning drafting at Chabot College, California, in 1980).

Lawrence Livermore National Laboratory My first job after getting my master's degree in mechanical engineering (from the University of Arizona, 1977) was in the Material Fabrication Group. Our mission to design and build a lathe capable of 0.000001 inch accuracy provided great experience. Additional experience in the design of experiments for laboratory programs proved invaluable for future work.

Intel Corporation This is my first experience with the design of computer housings for "silicon valley." Intel was just expanding beyond chips and printed circuit board products, and they had plans to get into building computer systems. I'm on the cover of "Intel News" shown trying to measure fan noise of a prototype "tower computer" in 1982. I was very fortunate to experience the tremendous work ethic in the culture at Intel.

Sytek Corporation This is my first design position where I had the responsibility of taking my own design thru all the stages of prototyping, tooling, testing, and manufacturing delivery. We installed our first CAD systems (CADAM) in 1986. This is the first company where I assumed some management responsibility and thus saw a more complete picture of where (mechanical) product design fits into a larger picture of the entire product delivery process.

Trimble Navigation Limited ("Trimble") This is my first designer position where I designed and documented with a CAD system (AutoCAD). My work here spanned nearly 25 years from 1988 thru 2013. We installed our first 3D CAD system (ProEngineer) in 1992. Trimble's products (portable GPS instruments) are intended to work in outdoor environments, and their customers need these products to function under adverse conditions (shock, vibration, temperature extremes, rain). I spent my early years in the Marine Division, moved on to the Survey Division, and became manager of a Mechanical Engineering Group in 1995. While manager, I was responsible for budgeting, resourcing, prioritization, design, and documentation. I also served as (overall) project manager on a number of projects and thus experienced how the "mechanical piece" fits into the overall design of products.

All of the above corporations had *different* accepted working processes for overall product design and *different* corporate cultures to operate in. It's hoped that most of the words written here serve as a valuable contribution to the reader's design process regardless of the corporate cultures they are in. The above work resume certainly spans a history of both how computers came to be used in the design process and how the industries evolved in their design processes. The design tools and products made for the marketplace will continue to evolve in the future, but we will still need basic mechanical design fundamentals to positively affect the overall product design process.

Continuing the Dialogue

I'd enjoy hearing from the reader to discuss any point in the text. It is hoped that readers can make suggestions to improve the text for any later editions. To that end, I can be contacted at TSerksnisDesign@gmail.com.

Contents

Chapter 1
Successful Design

This chapter gives you an introduction to designing enclosures for electronic products and defines a "successful design."

We'll discuss the designer's role in the setting of product requirements, where the designer fits into the overall product development picture, the importance of communication, and the initial factors to consider when beginning a design.

Before we get started, let's briefly define what we mean when we talk about an "electronic product." It is a product that has a circuit board in it and usually has some input/output device such as an LCD. Examples of electronic products include cell phones, digital cameras, and the ultrasonic toothbrush.

An electronic product enclosure is the item that surrounds and supports the circuit board. The enclosure is what makes the device usable to the consumer. The enclosure is necessary for a number of reasons – to protect the electronics (the circuit board and LCD) from the environment or from a physical jolt (such as dropping the product). The enclosure provides access to input information to the device, via keys or buttons perhaps, and allows information to be transferred from the device. The enclosure provides structure so that the circuit board logic is supported and protected.

Examples of some very effective product enclosures that have been developed in recent years are the Apple iPhone 7 or the HP Spectre laptop computer (both, circa 2016).

In essence, a successful design of an enclosure will be the one in which the design has conformed to the product's written specification (spec) and has been done within the cost and time parameters that were set. Let's now begin our exploration of the process of designing these enclosures.

1.1   Design Guide

This text is intended to place in a single reference, a *design guide* for the *successful mechanical* design of an *electronic product enclosure*.

© Springer International Publishing AG, part of Springer Nature 2019

T. Serksnis, *Designing Electronic Product Enclosures*,

https://doi.org/10.1007/978-3-319-69395-8_1

Let's break down some of the words of the above sentence for further definition (with the word "successful" defined in its own subtopic).

Design Guide

This text is a starting point, a point of reference. The designer will be using many guides in their work; this text is intended to be a general help and serves to augment the designer's entire past experience and their present organization's established processes.

Electronic Product Enclosure (EPE = Electronic Product Enclosure)

The electronic product enclosure consists of both the external and internal structural elements of a product. It includes any of the hardware used for user interfacing, any of the connectors used to interface cables, and any elements that the user will physically feel and see. Many electronic enclosures contain one or more PCBA (Printed Circuit Board Assemblies), and these must be protected against the rigors of normal usage.

An enclosure could be very simple or be extremely complicated with thousands of separate parts. One of the designer's first tasks will be to define the "system" that they are designing, and that is covered in a later chapter. The term "enclosure" (in this text) will be on the less complicated end of the spectrum, and the methodology explained can be extended into the more complicated design situations.

The EPE Designer

This is the person responsible for the design of the enclosure for an electronic product. In many cases, it is a mechanical engineer, but it can be someone with a background in mechanical engineering or who has the experience of the discipline. A good EPE Designer will have the following characteristics:

- Ability to understand and conform to the product specification
- Be able to add to and help create the product specification
- Create inventive solutions to the problems presented by the product

Thus, the EPE Designer must be able to both be creative and still follow the major objectives of the project.

1.2 Defining the Overall Team

The intent of this section is to show that engineering (and mechanical engineering in particular) doesn't design products by themselves; they are certainly a part of a team. Characteristics of the overall team are that the team can be:

- Of a small or large size
- Located in one location or distributed worldwide
- Limited in resources or have access to almost unlimited resources
- In possession of the latest tools, or not
- Motivated by a variety of reasons for accomplishing their goal

- Varying in experience

The entire engineering effort consists of an amalgam of design among several disciplines. These disciplines include:

- Electrical engineering
- Software and firmware engineering
- Mechanical engineering (including structural and thermal)
- Industrial engineering
- System engineering

Therefore, it is recognized that mechanical engineering is only a part of the overall engineering design of an electronic product, and many of the decisions made are in cooperation with the other disciplines. Contemporary product design should balance various trade-offs among all of the factors that go into the production released product.

Indeed, the *entire* engineering effort (all of the disciplines from Sect. 1.2) is only a part of the *overall* effort that goes into the release (sale) of a product.

Besides the engineering effort, contributions result from the following groups:

Each group is defined, followed by how specifically the mechanical design "interacts" with that group. All of this is meant to emphasize that the mechanical design is not done "in a vacuum" but rather as part of a multitasked product delivery team.

Marketing (Including Input from Sales) This organization is responsible for the product definition, that is, defining what the customer wants and what the product will be from the customer viewpoint. This "product definition" usually takes the form of a document that engineering will accept as the product requirements. Marketing also has the responsibility of overseeing how a particular product will fit into the overall product line of the company (or division of the company).

The EPE Designer interacts with Marketing in the effort to define how the product will function, how that functionality will present itself to the customer (user interface), and how the product will look to the customer (industrial design).

Operations (Manufacturing) This organization is responsible for the complete flow of materials for individual components and how those individual components get fabricated, assembled, and delivered to the customer. If engineering's responsibility is to produce the product documentation, operations should be able to take that documentation and get that product produced that meets the product specifications.

The EPE Designer intersects with operations by making decisions on part fabrication techniques, vendor (supplier) selection, and any trade-offs between quality/cost/appearance.

Testing (Design Verification) This organization is responsible for testing both the prototyping and mature designs. This can be accomplished by resources within the mechanical design group (itself) or by an independent group setup for this particular function.

The EPE Designer intersects with the test function by either conducting or reviewing test results. The testing done on the product is actually a part of the product requirements document (PRD) and that it must be proven that the product passes testing as defined in that document. For example, if the PRD states that a product must survive a one meter drop, then a test must be defined that states considerations such as:

- How many drops of a single item (under test)
- Impact faces or corners of that item
- Environment that testing is to take place (such as ambient temperature)
- Statistical concerns (such as how many single items must pass testing)
- Order of testing (among various tests that unit will undergo)
- Definition of "survive" (degree of functionality or appearance after test)

Quality Control/Quality Assurance This organization determines whether the acceptability limits of the individual parts (or entire assemblies) meet the standards both specified in the individual product specification (the drawing) and in the established overall corporate standards. Quality control would be concerned with tactical situations, while (corporate) quality assurance would be more concerned with strategic situations. Most companies have various ways of both controlling and monitoring the quality of the product and certainly get involved with customer satisfaction and service issues.

The EPE Designer intersects with this organization by specifying on their documentation the acceptability limits of each part and can go all up to include assemblies. Typically, acceptability limits take the form of:

- Size (geometry) control as specified in drawing tolerances
- Material and plating specifications stated on drawing
- Cosmetic flaw rejection criteria stated on drawing
- Functional specification as stated on drawing
- Determining the "critical" nature of some aspect of the part documentation.

Service This organization is responsible for the repairing, warranty, and return of product functions. They help determine course of action for field problems with the equipment.

The EPE Designer intersects with this organization by designing-in a reasonable process for the disassembly and repair of the product. Of course, a design with a designed-in high reliability will have less reason to repair. It's also possible to provide for methodology to determine misuse of the product.

Project Management This organization is responsible for tracking the project for:

- Time allocation – meeting deadlines that are committed
- Resource allocation

- Priority management (for a single project and relative to projects competing for the same resources)
- Compliance to specifications for the product
- Meeting cost goals
- Reporting status of project

The EPE Designer intersects with this organization by reporting estimates of time and resources for all separate line items of the mechanical part responsibility. This starts with product conceptualization, design, prototyping, and testing and continues on into final release documentation. Estimates of time and resources are updated as milestones are met.

Upper Management Included in this group is anyone who is responsible for the project and has a need to understand the project. Project updates would be provided to this group at specific times during the project. Upper management would provide leadership and vision to the project.

The EPE Designer intersects with upper management in an indirect manner. Reporting of project status is relevant at any time and is usually provided thru the project manager.

1.3   Product Requirements

Determining success is a matter of meeting (or exceeding) the requirements of the project. This is a simple statement but is actually very complicated in its interrelated aspects.

A project could be determined successful if it met its goals. These goals can be addressed in (one or more of) the following written documents.

Product Requirements Document (PRD) This document can go by a variety of names (it will vary by company). Basically, it is a "contract" of sorts that attempts to specify the basic functionality of the product. It can be as simple as a few paragraphs or extremely complicated. It can contain:

(a) A description of what the product will accomplish for the customer – it usually does *not* specify exactly how the product will work. That is, details on "how to get there, from here" are not explicit. This description uses words on the "final outside appearance" of the product rather than the details of the "inner workings." Follow-on documents (or specifications) can also specify details of the product. Again, the PRD forms an agreement between marketing and engineering as to what the product will be. The PRD can vary in its content detail. It *is* (should be) updated, during the course of the project, as elements get revised or added to. At each overall product review, it should be compared on the extent of how the design is conforming to the PRD.

(b) A description of how the product will interface with the customer. This would include:

- How information is displayed to the customer or how the information will get from the customer, to the product. This can be visual, auditory, or tactile.
- Various interfaces to the product, such as connectors, switches, or buttons.
- Labeling or icons intended to provide information to the customer.

(c) A description of the various components of the product. That is, if the product (the product being designed) needs additional equipment or cables to function in a larger system, then a description of the various parts of the "system" will need to be described. Thus, one will need to "draw a boundary" around *exactly* what *this* product (being designed) is. What exactly is the "deliverable" to the customer?

(d) Indication of the final aesthetic (visual appearance) of the product. Colors, textures, and industrial design are usually very well-specified.

(e) A listing of the environments that the product will both operate and be stored in. This includes temperature, shock, drop, vibration, humidity, water egress protection, shipping conditions, altitude, and specific corrosive atmospheres.

(f) A listing of any standards that the product will need to pass. This includes both safety and regulatory standards such as Underwriters Laboratory (UL) for safety, federal communication compliance (FCC) for electromotive magnetic interference (EMI), and the (literally) hundreds of other compliance standards that are a real part of today's design world. Some of these standards are country specific, while others are accepted on a worldwide basis. Obviously, anything to do with medical, food, or children's toys will have their own rigorous testing standards to pass.

Internal Test Reports These indicate positive test results. These are the results of testing done to show that the requirements as set forth in the PRD have been passed. If the tests haven't been passed, then there are action plans initiated to improve the product and conduct further testing.

Reports from Initial Customers This is "alpha" or "beta" testing where customer feedback is positive or negative. It is hoped that customers are gaining measureable value from the product. Reasonable improvements to the product can be made when this "real-world" feedback is available. "Alpha" testing is usually done with in-house personnel who are simulating the actual customer, while "beta" testing is usually done with existing customers before shipment to actual (paying) customers.

Project Management Reports

(a) On expenses (expected vs. actual). This includes expenses for salaries, capital equipment, tooling, etc. Monitoring of expenses can lead to analysis of the true "payback periods" of the project and better predictions on expenses for future projects.

(b) Status on milestone dates (expected vs. actual): as with expenses, monitoring of how well the project achieved its time commitments leads to an indication of the true "payback period" of the project. Analyzing where milestones were not met can lead to better predictions for future projects.

Ongoing analysis of "success" (as the product matures in the field) can be measured by:

Quality Assurance Reports   These contain information about customer satisfaction and warranty returns: any issues or problems with the product must be *quickly* addressed so as to protect the company's reputation in the industry. If revisions need to be made, they must be implemented with great urgency. Thus, if customer satisfaction reaches some set level of reliability, the product design team will have achieved success.

Analysis of "Lessons Learned"   From all disciplines on the project: every project will contain items where things could have been done better. Continuous improvement should be strived for. There should be a way to gather feedback from everyone in the product design process on what items would need to be improved. This will enhance the success rate of future projects. More on this subject is presented in Chap. 13.

Sales   Expected vs. actual. Sales figures can indicate the success of the project – in the sense that marketing has predicted the need for the product, engineering/operations has delivered that product to the customer, and the customer does (indeed) value that product. Or, in the opposite case, sales can be less than expected (predicted). This could have happened for a variety of reasons (such as):

- Product is not (exactly) what the customer needed (price too high/performance features too low).
- Product is too late out into the market, that is, it took too long to get the product out into the market, and the customers now have better choices.
- Product is too early into the market (not enough "early adopters"). This happens when the technology of the product doesn't match what customers (at the time) value or other supporting technology isn't available as yet that would make this particular product fully useful.
- Low reliability.

All of the above reasons should be placed in the "competitive arena." That is, most products have competition in their markets. Customers will choose purchases based on their needs for performance, price, and quality. New technology solutions must compete against the older solutions.

It would be rare to have all of the data available at product release to determine how "successful" the product design effort is. Product design usually has increased risk of success if:

- Milestone completion dates are unreasonably shortened.
- The design has a high content of brand new components.
- Changes (additions) to the project occur at an unmanageable rate.

Successful design has been simply described as:

1. Function to specification
2. Delivery on time to project schedule
3. Delivery at predicted costs

Of course, projects can *exceed* functionality, be delivered *ahead* of time, and perhaps be even at a *lower* cost. This would be cause for celebration (although some examination needs to go into why "actuals" didn't match "predictables").

Behind the above "simple statements" for successful design is however some very large implications and that they are *not* so "simple." Let me break down the above three variables a bit. All three are interrelated on several levels.

1.3.1 Function to Specification

Specifications take many forms. They can be written documents, notes from a meeting, or even verbal instructions. The way that projects create specifications varies from company to company and indeed can vary within a company itself. Also, you, the particular Designer, can come in at various stages in an overall project. Therefore, there is no particular way that the work description can manifest itself to you, the EPE Designer.

Although the EPE Designer is not ultimately responsible for setting the full product requirements (in the specification), the designer's input is *critical*. The EPE Designer will be tasked with providing input as to just how far the limits of the design can go. For example, if the Product Requirements "arbitrarily" determine that the shock levels for the product are 40 g maximum, the EPE Designer must do some research (or some initial testing) as to exactly what shock level is possible or what levels have been achieved in the past. Therefore, the 40 g level is initially "proposed," and the EPE Designer must agree to that level or put forth arguments for a different level. It may even be possible that *higher* g levels can be agreed to. Similarly, if cost targets in the Specification seem overly aggressive, the EPE Designer must do some "homework" on their portion of the budget that provides reasonable data back to the project specification.

The important item to concern the EPE Designer is with the writing down of a specification and the later agreement (formal or informal) among the various members of the project. For example, let us say that the general task is one of designing a removable Disk Drive Module. Here are some possible scenarios that lead to its "successful design."

Task (Example): Removable Disk Drive Module <u>Scenario #1: Minimal Input (to the Designer) – The Beginnings of a Specification</u>

This would mean progressing with the design without much more than verbal information (as given above). The designer would likely proceed to find out items that would affect the design such as:

1. How many times is the drive to be removed? Will it be just for maintenance or is it more like once a day to secure the data?
2. How large can the module be designed?
3. Is there an existing opening for the module (in the base unit)?
4. Is there a shock concern for the disk drive (what levels of shock)?
5. Will this module be used in other base units?

These questions must be considered at this point to get some agreement as to how the design should proceed. Formal or informal meetings (communications) should now be held to get the answers, even if the present answer is "unknown at this time."

The important item to do at this point is for the designer to create his/her own "working spec" for the design by writing down what *is* known (and unknown) about the design. This document (again, you are creating the specification) can now be revised as often as necessary, each time with agreement of those persons concerned with the project (at this time). The document at this point does not have to be of any great length. It can be as terse as necessary, for example, in the example of the disk drive that we have started:

Project: Disk Drive Module
Author: (the designer)
Revision Level: 1 (date)

- Design requirements:

  - Disk Drive Module will function under a shock load of 20 g.
  - Disk Drive Module will survive a shock load of 100 g (nonoperating).
  - Ambient air near the Disk Drive Module will be 30 °C maximum (operating).

The above is just the start of the specification, but as more is known (specified), the designer can proceed. The design can proceed because the designer now has some idea when they have been successful, that is, if the design passes testing designed to determine whether or not the design has passed the specification. In this "Scenario," minimal input, the specification will certainly be added to, and many people involved with the project will need to review and approve the specification. However, the designer can, at least, proceed to make some progress or show some design options.

- **<u>Scenario #2: A Complete Specification</u>**

This specification describes *in detail* all of the requirements of the mechanical design. (The specification actually describes in detail all of the requirements of *all* design elements of the design, not only the mechanical portion, but we'll concentrate here on the *mechanical* requirements.) It would include:

- Product description
- Product financials
- Product scheduling

It would include in its design requirements such needed detail as:

- Module to plug directly into backplane for power and signal requirements
- Module to slide on a nonmetallic surface for ease of entry/exit

(Plus a whole host of other requirements, including, environmental, ergonomic, electrical interfaces, agency approvals, testing required, etc.)

- **<u>Scenario #3: A Working Specification</u>**

This specification is (by far) the most common specification that the designer is adhering to. The specification's completeness is somewhere between the "Complete" and the "Nonexistent" (beginning) specifications (Scenarios 1 and 2). With the working specification, the project manager usually has some idea of the design constraints, but all aspects have not fully been vetted out. The specification is now under "change control," that is, it is being updated fairly often in the beginning phases of the project, and any changes or additions are being reviewed by the project personnel with signature responsibility. As the project matures, major changes are under extremely tight scrutiny as these changes can greatly affect project completion dates and milestones along the way.

1.3.2 Delivery on Time to Project Schedule

Various schedules are prepared during a project. Each (approved by the project team) schedule is a "snapshot" of what the current project schedule is to be. The first schedule of importance would be the schedule that is used to justify the project. This schedule would be the one that is being used to be the "net present value" (NPV) of the project. This NPV project schedule would include best estimates of:

- Person resources needed to finish the project (by a given date)
- Capital resources needed to finish the project (by a given date)
- Expected sales of product and at what price sold (if product sold by a given date)
- Expected cost of sales of product (if product sold by a given date)

Thus, there is an expected "value" that the project has if that project is completed by its expected date which is the date that the schedule presently states.

However, various considerations can change during the course of a project. They can be:

- Technical issue arises that changes implementation of original design.
- Personnel working on project change (either particular members or the size of team).
- Scope of project is revised (either increased or decreased).

- Revisions to costing of various project components as project proceeds.

As each (above) consideration changes, the project team will meet to determine its effect on the overall project schedule and see how it affects the NPV of the project. A determination is made to either continue the project with the revised NPV or discontinue the project. Usually, as the length of a project is extended, there would be more expenses associated with that extension, and the NPV would be lowered.

Getting back to determining "success" as "delivering on-time to project schedule," each project can analyze if they have indeed delivered that project "on-time to project schedule" by dissecting the causes for any extension of the time schedule. If the extensions are deemed "reasonable" and "justified," then the project could be considered to be a success in this regard.

1.3.3 Delivery at Predicted Costs

Just as the schedule can change (in Sect. 1.3.2, above), the costs of either the product or the costs needed to design and deliver that product can change. Costs can change due to the following:

- Personnel resources needed to finish the project are revised.
- Capital resources needed to finish the project are revised.
- Expected cost of sales of product are revised.

So, just as the schedule changes and it is determined whether the changes are "reasonable" and "justified," the project costs could change in a similar manner. If the changes to the project costs are determined "reasonable and "justified," then the project could be considered successful from a cost standpoint.

Certainly, all three of the above factors (specification, time, and cost) need analysis during and after the project. Chapter 13 (Continuous Improvement) explores further aspects of determining whether the project can be considered "successful."

More information on product costing is presented in Sect. 1.7 on Engineering Economy.

1.4 Sketching Versus Detailing

An EPE Designer must know when to shift between either of these modes:

- **Sketching** or brainstorming: this is a very quick ideation phase. It is usually done with a pencil (don't use an eraser; that will slow the thought process). Nothing is detailed – it all seems to fit perfectly on these sketches. Scale isn't really important; that will come a bit later. Feedback from others is attained. Speed is the main focus here; the designer is getting major choices on paper so that plusses and minuses of several choices can be decided upon. What are the other choices for orientation? Up? Down? Sideways?

- **Detailing**: that is, providing "some amount" of detail. The amount of detail needed is dependant on the criticality or the uniqueness of the situation. In the sketch (phase), *everything* works, you have "glossed over" the items that may be *stumbling* blocks. You have done that to *speed* the overall design process, but now in the detail mode, more critical analysis is needed. While in "detail mode," you work out (more) exactly some critical parts of the design, the parts that a designer recognizes as "deal breakers." Details usually need CAD design to provide real geometry and scale to the situation. Again, design reviews can be critical to continue quickly down agreed-to design paths.

I'd like to continue the discussion on basic layout, with the assumption that we are working on a brand new design. Many of the concepts will be applicable to the continuation or modification designs. Also, I'd like to proceed with the design discussion as if we are still in the "sketch phase."

1.5 Design Reviews

Along with going back and forth between the "sketching" and "detailing" modes (above), the designer needs a solid feel as to when to get others on the design team to review or comment on their designs (in whatever phase the design is in). Some of these design reviews are *very* formal, while other design reviews can be very informal.

Formal design reviews are usually done to a scheduled milestone on the project schedule. They include specific members of the design team and have definite sign-off by those members.

Informal design reviews happen sporadically and/or spontaneously. It can be as simple as the designer going over to the next office and asking a colleague to "take a look at this" or calling a short meeting among a few people that the designer feels are close enough to the design issues or have previous experience with similar designs.

Some General Comments About Design Reviews:

1. Take attendance, and note who is at the meeting.
2. Take (at least, cursorily) notes on all of the (relevant) issues that are raised.
3. Possibly invite someone to the design review, who is *not* familiar with the basic design. They could be someone from a different function or department within the company. Sometimes, this person can add a different "take" on the issues as they are viewing them from a different perspective.
4. Briefly review the main goals of the design.
5. Some ideas brought up will already have been thought out by the designer; that's fine, and just go quickly thru your rationale.

6. Some ideas may seem (initially) to be not valuable or not "on point." Just note them and move on; it's possible that you may see the value at a later time.
7. You *will* get the value out of the design review – it always happens if you are "open" to it. Thank all of those involved as they have given you their insight and experience.
8. *Publish* your notes on the meeting to those that were in attendance and to the entire project team. This will log further action to be taken by you and others to make progress with the design. Ask for further comments from your Design Review Team.

1.6   Communication

Communication is a necessity for a design to be successful. That sentence stands by itself but is worthy of a more full discussion.

1.6.1   Purpose of Communication

The purpose of communication is to convey information about the design. This is required as that information supplies answers to questions, documents the design as it presently exists, and documents the evolution of the design. Communication can be written, filmed, or verbal. Written documentation takes the form of:

- Specifications
- Drawings
- Project meeting notes/schedules
- Notes in notebooks
- Emails specific to a project or program (any digital communication)

Film documentation are camera recordings of project proceedings, tests, and events. Verbal communications are any of the words spoken to move the project into a completed state. All important verbal communication needs to be put into written form so that all members of the project can review those communications.

1.6.2   Value of Communication

Great communication will make a project more successful. Great communication has the following attributes:

- Accurate – the information is true and backed by testing/documentation.
- Concise – the information is straight to the point without excessive words.

- Distributed – to all those that need the information.
- Speedy – the information is quickly disseminated.
- Offers solutions – proposing a solution to problems invigorates the solution process.

1.6.3   Links in the Communication Chain

One of the goals of communication is to get that information to the people who need the information. As seen in Sect. 1.2, the project team can include a lot of disciplines. Email distributions are easy to create along with document control distributions. The EPE Designer should decide among all of the people involved on the project team, who is critical to making the decisions that need to be made and who is copied for status purposes only.

1.7   Engineering Economy

No chapter on successful design could be completed without a discussion of the basic principles of Engineering Economy. "What something costs" is a paramount consideration in just about any endeavor. In many product designs, it will be one of the main causes of product success or failure. Chapter 4 will include a discussion on the trade-offs between cost vs. time vs. specification, but for now, let's begin the discussion stating that the EPE Designer must be capable of coming up with cost information and some break-even analyses. In Ref. [1], it is stated that "If an engineering project is to succeed in meeting human needs, it must be designed and operated in a way that promises both physical and economic feasibility."

As cost is so important, the designer needs to:

1. Be aware of what the cost goals (for both the individual part and the complete assembly) are for the design.
2. Be aware of what the material and process options are for the part being designed. It shall be the responsibility of the EPE Designer to present various options for attaining (or reducing) cost targets by (perhaps) compromises on functionality. In no cases should the designer ever compromise on any safety considerations. All options must be clearly presented to the management so that any trade-offs can be thoroughly determined. This is one of the most important creativity elements that the EPE Designer brings to the overall design of the product.
3. Propose solutions to materials and processes that are appropriate to where the product is in its overall life cycle. Certain solutions may be more appropriate to early production (where time to market is very critical) than in mature production.
4. Provide cost information, based on the appropriate quantities being ordered, back to the project team so that this important metric is always well-known.

Let's get into an example. As shown in Chap. 4, the choice on whether to "tool" a part is mentioned. This type of problem can be analyzed by choosing between: (200 parts per month needed in each case)

- Choice A: un-tooled part cost is $5.00 from Vendor A
- Choice B: design a tooled part and have Vendor T make the tool. Estimate of costs: tooled part cost is $1.00 and tooling cost is $4000.00

At what time period will the tooling cost and new part cost be *equal* to the old part cost? This is what is known as the "break-even point."

Answer: this is easily calculated or graphed (see Fig. 1.1):

$$(\$5 \, / \, \text{part} \times {}^{"}\text{M}^{"} \, \text{months} \times 200 \, \text{parts} \, / \text{month}) = (\$1 \, / \, \text{part} \times {}^{"}\text{M}^{"} \, \text{months} \times 200 \, \text{parts} \, / \text{month}) + \$4000 \, .$$

"M" is shown to be 5 months. Thus, it will take 5 months for Choice B (tooled version) to "break even" with Choice A (the un-tooled version).

The total cost at 5 months is either:

$$\text{Choice A} = \$5 \, / \, \text{part} \times 200 \, \text{parts} \, / \, \text{month} \times 5 \, \text{month} = \$5000$$

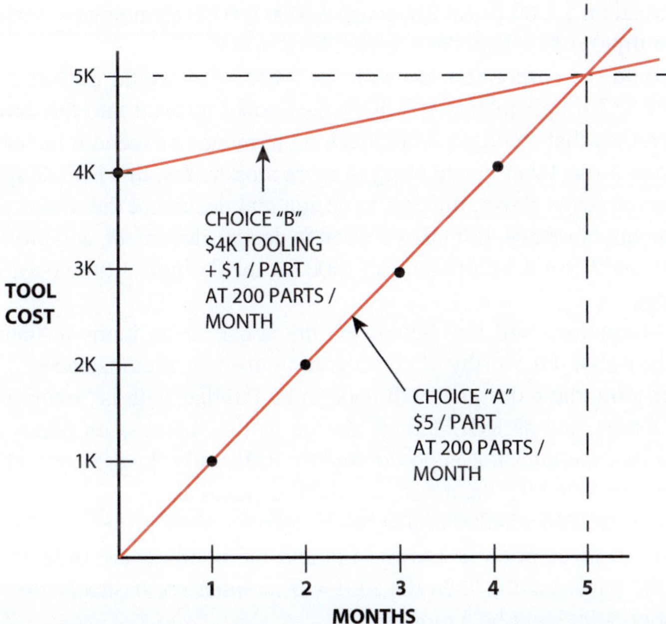

Fig. 1.1 Tooled vs. Untooled Part Break-even Point

$$\text{Choice B} = (\$1 \,/\, \text{part} \times 200 \,\text{parts} \,/\, \text{month} \times 5 \,\text{month}) + \$4000 = \$5000$$

After 5 months, the total cost will be less for Choice B, the *tooled* part. Clearly, if the product is produced for 5 months or more, we would choose to tool the part.

However, the above "economic reality" actually has some complications. Choice B costs (above) did not include:

1. The cost to design and document the *tooled* part.
2. The cost of bidding the tool and deciding that Vendor T was the best tool vendor.
3. The cost of prototyping the tooled part (before approving the drawing of tooled part).
4. The cost of getting a first article of the tooled part approved (we'll assume here that the initial first article is approved – hopefully, no tool modification is required as this would be an additional cost).
5. The cost of testing the tooled part (as a replacement for the un-tooled part).
6. The cost of "using up" the un-tooled part and switching over to the tooled part. (Should the assembly where this part is located be date coded to note changeover?)
7. The cost of the money for the tooling. That is, the $4000 cost of the tooling is actually worth more than $4000. Let me explain.

If the $4000 was *not* going to be given to the tool maker vendor, it would be an earning interest for the corporation. A "simple interest" calculation would have the interest earned on $4000 be (at 2%/year): $4000 \times 0.02/12$ months $\times 5$ months (the break-even time) = $33. But, the question here may be:

What could the corporation do with the $4000 that would be better than just creating the $33 simple interest? Perhaps they could invest it into the development of a new product that could generate much more money or spend it on another tool on a different project that would bring in more money. Yes, this isn't simple.

In the seven items above, the cost to design and document the *tooled* part is not a trivial amount of money. If it takes 1 week to design, document, and prototype, the tooled part could cost the corporation $1000 for the designer's time (say, at $1000/ week salary).

Most corporations will not "factor in" the above seven items in their "break-even" analysis, but it is worthy of consideration in some circumstances.

Another term that a designer will need to be familiar with is "return on investment" or abbreviated as ROI. This is similar to the "break-even point" as stated earlier, but that the question is posed slightly differently. In the same problem as above (Choices A and B), the question would be posed as:

What is the ROI on a $4000 tool to bring down the piece price of a $5 part? Once the new tooled piece price is known ($1) and the numbers are ordered (say, per month = 200 parts/month), then the ROI = 5 months. So, basically the return on investment of $4000 will be 5 months.

Chapter Summary

Chapter 1 gave us an introduction to designing enclosures for electronic products. As we are not alone in this endeavor, the chapter also defined the other major groups involved in the design and what their function usually is. The chapter introduces us as to how the design will be deemed "successful" and how the design will meet (or exceed) the (defined) product requirements.

This chapter takes us thru several "design scenarios" where we get everything from a "fully defined" specification to specifications that are minimalistic.

We defined the way to move along designs in either a "brainstorming" or "highly detailed" mode. The needs of both Design Reviews and good communication paths in a general design process were discussed.

Finally, the subject of Engineering Economy was started. This will be further amplified in Chap. 4.

Reference

1. Thuesen HG, Fabrycky WJ, Thuesen GJ. Engineering Economy, 1971, Prentice-Hall Inc., Englewood Cliffs, N.J.

Chapter 2
Building the Design

Our designs will all start with just an idea for a product. Those ideas will need to be proven, so, we'll move on building prototypes, and if those prototypes seem to work via testing to some written specification, we'll document our designs via drawings. We'll need this documentation to be able to build more products in a repeatable manner.

This chapter will take us from the point of just having an idea for a product, all the way to optimally placing all of the individual objects that will make up the final *working* design.

We start out our design with a "blank paper," and that paper will get filled with physical objects. First item of concern will be to determine if the paper is indeed "blank" or are there some beginning constraints. Next item of concern will be to determine exactly what physical objects are to be included. We should then assume that there is an *optimal* placement of those objects based on the objectives of the overall design, so we'll conclude with some words on the choices we have on that object placement.

2.1 Beginning Point

Designers are tasked with either continuing work on existing designs or starting a brand new design. Let's spend a little time seeing what the differences are with those starting points.

- Brand new design: This is a "clean sheet" start for the designer; they would basically have no constraints, other than complying with the specification. We'll have an entire section on what exactly a specification is and its various components.
- Continuation (or adding to) an existing design: This is a variation on the brand new design, but only a *small* part of an existing design is to be modified. The designer here has many of the same challenges of the brand new design, but the

© Springer International Publishing AG, part of Springer Nature 2019
T. Serksnis, *Designing Electronic Product Enclosures*,
https://doi.org/10.1007/978-3-319-69395-8_2

additional work must utilize the existing design. We will have a separate section on defining what exactly the "system" is in this context.

- Major modification of an existing design: Again, this is a variation on the brand new design, but in this case a large part of the original design is to be modified. The designer here is tasked with changing a part of the overall design, so there will be more constraints than a brand new design.

So, it's important to know where the present design effort will fit into what has been previously done. Our "basic layout" can proceed either with or without the constraints of previous work.

2.2 Defining the Design Boundary: System Description

A few words on defining the "system" being designed: Designs can be extremely complex and large (think space shuttle or large passenger jet), smaller systems (think automobiles), or yet smaller systems such as a personal computer, coffee maker, or cell phone. The scope, cost, time, number of resources committed, and interfaces to other equipment vary with all of these design projects. It is imperative that the designer keep in mind the system being designed. This is important for a variety of reasons, a few of which are:

- Focus on the personal responsibility (scope of work)
- Awareness of other equipment that must interface to this design
- Overall "system" functionality (not just the function of the subsystem)

Even something as "small" as a cell phone functions as part of a larger system. That is, the box that the consumer purchases can contain:

- Cell phone
- Battery charger
- Cables
- Sim card
- Instruction manual
- Other shipping materials (labels, bags, bubble wrap)

(We'll limit the discussion of the "system" here, as one can even think of a larger system that would include the cell phone towers and satellite systems.)

We start here at system description because most electronic enclosures surround and support a product. Sometimes, one product can be thought of as part of a larger product. For example, a network adapter card (a product itself) can be placed into a microcomputer (a second product) and form an entirely new product, in this example, a networkable microcomputer. Things get even more complicated as the networkable microcomputer itself forms a part of the network which may be an even larger product.

Looking at things in another way, we may be tasked to design just a subsystem of a much larger system. Thus, our "system" may be not even a product but just part of a larger "system" that has been broken down into (time) manageable pieces. For

Fig. 2.1 System
description

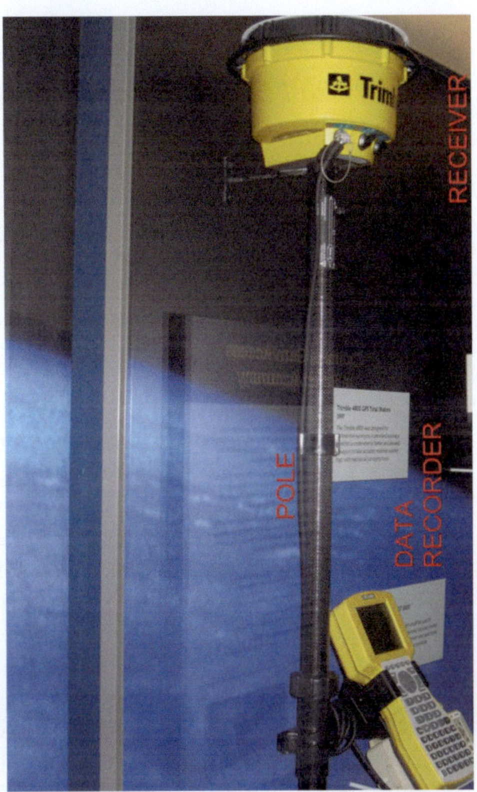

example, we could be tasked to design a data recorder as part of a (larger) surveying system. This system is shown in Fig. 2.1. This system consists of (at least) three major subsystems:

- Data recorder
- Data recorder mounting bracket
- Survey pole (labeled "pole"), which includes another subsystem, the data recorder bracket

There would actually be even more individual portions of this "system" including cables, shipping container (box), and instruction manual (but let's neglect these for this example). As a side note, the system shown in Fig. 2.1 is a photograph of a Trimble Survey system appearing in the Smithsonian Museum in Washington, D.C.

We have become a team member of the "surveyor system design group" and will be designing a portion of overall design (the data recorder portion).

Therefore, the first task for us is to determine (specify) what exactly we are to design (sort of like "building a fence" around the project that we will be responsible for in a given amount of time). To accomplish this task, we will need a specification. (Refer back to Chap. 1 discussion on specifications.)

2.3 The Design Process

2.3.1 Overall Project Start to Project Finish

Designs can proceed in *any* number of ways. All companies vary in how they execute the entire product design process, but they do have some common characteristics. No particular way is absolutely correct; it is the end result (specification conformance) that is a measure of success. A design usually proceeds as:

The EPE Designer will have a "lion's share" of the responsibility of the following tasks. They will be both a "doer" and an "enabler" of much of what is done. If they don't do the work themselves, they certainly are responsible for the work.

1. Sketch of idea – this is the "ideation" stage of the project. Words must be turned into a picture representation of those words. Once the idea takes some form, it can easily be reviewed and revised. Some of the people in the review group need a "picture" of the idea to really see what is being proposed.
2. Review of idea and authorization to proceed to prototype – this action takes the "picture of the idea" and turns it into something the team can actually touch. What seemed fine in sketch form can now be picked up, held, and used in a way a customer will use the product; the prototype is a full-scale, three-dimensional picture. The "authorization to proceed" is important in that projects that are usually limited in time and money, so these expenditures must be agreed to by the team. Steps 3 and 4 (below) actually create the prototype.
3. Drawing (file creation) of idea for prototype fabrication – usually, a sketch is turned into a digitized drawing file that allows the design to be fabricated.
    (Design will now be at Revision 1.) Italics are included to show "revision level" of the formal documentation, which is further expanded on in Chap.12.
4. Prototype fabrication (physical parts) – the project team will determine the cost and time constraints of producing the prototype. Sometimes, just a "quick-and-dirty" prototype is needed to make good progress; sometimes a prototype constructed to exacting specifications is needed. The EPE Designer should have very good perspective on what is needed for this stage of development.
5. Prototype analysis and testing – once the prototype is received by the team, it is tested to see how the prototype conforms to the specification. The project team determines just exactly what testing needs to be done to make the decision on how to proceed after that testing.
6. Review of prototype and test result – test results are reviewed by the team, and revisions are proposed.
    (Assuming Revision 1 needs improvement, we'll revise design to Revision 2.)
7. Change to improve prototype (drawing and prototype) – this is the start of an iterative process which will finally result in the design conforming to the product specification.
8. Further analysis and testing of Revision 2
    (Assuming Revision 2 conforms to product specification.)
9. Final documentation produced/final testing/final review
10. Formal approval of design for Production Release

Note that "Production Release" in the above process allows the production of "some number" of units to be produced for sale to a customer or to serve as a larger number of units for a more expansive test program. Corporations can differ in a number of ways on their procedure for releasing and testing their products for their customers. Also note that most projects would have many more revisions than the two revisions shown, but the project generally proceeds as shown.

2.3.2   EPE Designer's Starting Considerations

There is no "absolutely correct" way to proceed with a design for the EPE Designer. Each case has its own unique best way to make visible, required progress. Sometimes, a prototype, being put together in a few days, can spark an incredible new product breakthrough to the market. In other cases, a systematic approach to laying out a few possible solutions, taking months to put together, may be the best path. That being said, the following outline should prove useful to designs at least as a starting point.

1. Determine the use and requirements of the solution not directly related to load. Some of the more important of these requirements are:

    (a) Environment – where will the product be used? Examples are office/outdoors/at altitude/on vehicles.
    (b) Temperature – what are the temperature extremes of the environment?
    (c) Expected life – one usage, years of warranty, service?
    (d) Cost requirements – always an important consideration. Will definitely depend on number of units being produced and tooling budget.
    (e) Finish requirements – cosmetic details can greatly affect cost.
    (f) Size and weight limitations – what are the bounds of the present solutions in the industry? Affects materials/fabrication techniques chosen by designer.
    (g) Safety and regulation requirements – what are the effects of product failure?
        All of the above are very important considerations to be considered at the very beginning of the EPE Designer's design. For example, a different design results from indoor vs. outdoor environments. A different design results from a design expected to last "one time" vs. a design that needs to work after 1000 uses. A different design results from a design that needs to cost under \$5 vs. a design that needs to cost under \$100. By going thru each element above, the EPE Designer can determine some initial constraints.

2. Determine or estimate the working load from all the various possible types of loads that the individual member (and assembly) may be required to withstand. It is necessary to consider all likely combinations of loads and, if possible, to determine the relationship between load and time. Some of the possible load types are:

    (a) Static
    (b) Steady-state dynamic (vibration)
    (c) Transient dynamic
    (d) Impact or shock
    (e) Body contact, such as point loading or friction

(f) Other loading, such as thermal/gravity/acoustical

The above load determinations are also very important considerations for the EPE Designer's design. For example, a different design results from a 10 pound static load vs. a 100 pound static load. If those loads are changing over time, this will result in a different design solution. Determining the magnitudes and types of loads will directly determine the materials and cross-sectional shapes that are needed to support the electrical components.

3. Determine what the failure mechanism will be. Deformations occur due to loads being axial, shearing, bending, or torsional. Possible failure modes are:
   (a) General yielding (overall inelastic behavior)
   (b) Rupture or fracture
   (c) Sudden – caused by static or dynamic load on brittle material
   (d) Slow – caused by static load on ductile material
   (e) Progressive – caused by repeated load (fatigue)
   (f) Excessive deformation
   (g) Buckling
   (h) Creep – deformation under constant stress
   (i) Relaxation – changing stress under constant strain
   (j) Abrasion (wear)
   (k) Corrosion

By the EPE Designer determining how their design will fail (in its present state of design), it will be possible to revise that design to protect against that failure. Testing will also reveal some failure mechanisms. However, if some of these failure mechanisms can be thought of *before* testing, much savings of development cost can be saved.

In summary of the above three items, by determining the use cases, loading, and potential failure mechanisms of the design, the EPE Designer can proceed with the design with a solid base of understanding.

2.4   Optimal Object Placement

Most designs can be thought of as the physical placement of objects in space. The individual objects are the separate parts of the overall assembly. Some of the individual parts are completely known (they either are bought off the shelf from another company or are a reuse of parts previously designed in-house). Besides the "known" parts, other parts need to be completely designed new. These new parts can be produced in-house or completely specified to be produced by another company.

Electronic packaging design consists primarily of arranging subsystems into their most efficient arrangement. The first step in deciding upon this arrangement is to look at the separate volumes of the subsystems. These volumes, along with the "gaps" necessary between them for clearance, will generally set the "outer boundary" and, therefore, to a large degree set the overall size of the product. On occasion, the first criterion the designer starts with is the overall size of the product. From here, they must then decide if they can fit all of what is being asked for within the given overall size. That is, our subsystems may indeed be required to shrink to fit within this given overall size.

An aspect of the basic design process is shown in Fig. 2.2. This shows an object (in space) at some distance from an enclosure (shown as "the wall"). I'd like to start my discussion on the design for an electronic enclosure with a description of several

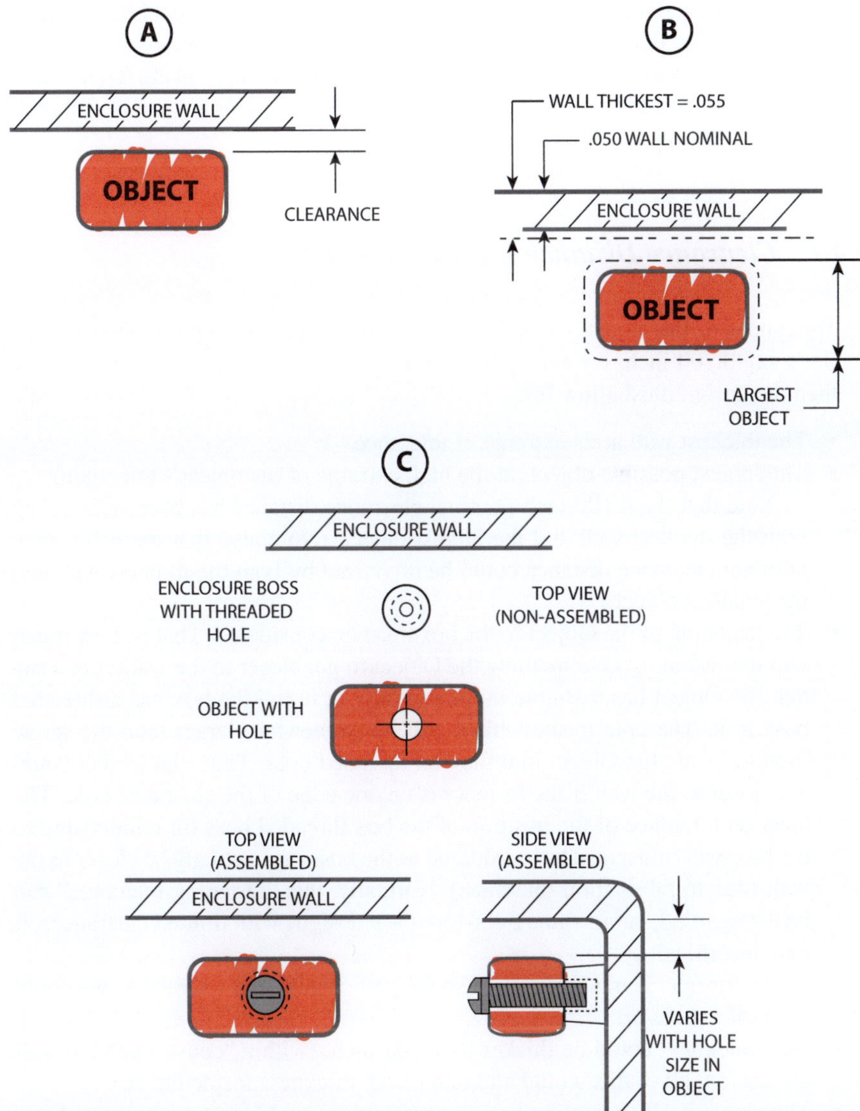

Fig. 2.2   Object/wall clearance

design "scenarios." Much of the discussion is for a 2D (plan view, from above) situation, but this is easily extended to include 3D (side view or the Z-direction), and I'll show some examples of that 3rd view:

Basic Object/Wall clearance: Fig. 2.2 shows an Object and a Wall. The "Object" could be considered just about anything. For example, it could be a printed circuit board assembly, an automobile engine, or any electronic component. The "Wall" can be considered the enclosure outside surface or the exterior of the item being designed. In just about every design, the designer has to determine the distance (clearance) between the "Object" and the "Wall." The idea here could be extended to determining the distance between Object 1 and Object 2. None of these determined clearances need to be the same as each other. The clearance in the X-Direction can be different than the clearance in the Y-Direction, which again can be different than the clearance in the Z-Direction.

2.4.1   Clearance Distance Is a Function Of

1. Tolerances of the object and wall: If one were to maintain a particular distance (let's say 0.100 inch, for example), and a nominal overall (outside) dimension, then the design must allow for:

   - The thickest wall at its extreme of tolerance
   - The largest possible object (at the high extreme of the object's tolerance)

     Note that the 0.100 inch nominal clearance distance has been *reduced* by both the thickest wall and the largest object. (Note also that the 0.100 inch nominal clearance distance could be *increased* by both the thinnest wall and the smallest object.)

   - The fastening of the Object to the box must be considered. That is, how much will the fastening system allow the Object to get closer to the wall? Let's say that the Object has a simple mounting hole in it and the box has a threaded boss in it. The hole in the Object will be (somewhat) larger than the screw used to fasten the Object into the box threaded boss. Thus, the Object could get closer to the wall if the fastener is on one edge of the clearance hole. The location tolerance of the position of the box threaded boss (in relationship to the box wall) must also be considered as the boss may actually be closer to the wall (due to fabrication tolerance). Normally, this "fastening tolerance" can be disregarded, but in some tight-for-space designs with limited clearances, it may be critical.

     So, for our clearance example so far, just considering tolerances, we could have tolerances of:

   - Wall thickness could be thicker by 0.005 inch. (with a "constrained" overall dimension, all of this would add to the wall thickness inside the box).
   - The object itself (mounting hole to object edge) could be at a maximum locational tolerance; this could be 0.010 inch.

- The mounting boss in the box could be closer to the wall due to its positional tolerance; this could be 0.005 inch.
- The mounting hole could be 0.010 inch larger than the (smallest) fastener diameter, allowing for 0.005 inch additional movement.

All of the above (4) tolerances add to 0.005 + 0.010 + 0.005 + 0.005 which would total to 0.025 inch.

Footnote: (Statistical note) Some designers would make a case for some statistical probability, less than 100%, that all (4) tolerances would go in one direction, and we would not (likely) have a total of 0.025 inch. Some conservative designers would assume that *all* tolerance will go in the "wrong" direction, and thus design is a "worst case." I'll generally disregard the "statistical" approach to tolerances for now, but it could be valuable in design situations where space is extremely constrained. See Sect. 4.8 for a discussion of:

- Tolerancing using sum of squares
- Tolerancing using Monte Carlo simulation

2. Movement of the Object relative to the Wall (during product operation): This is also known as "sway" clearance, that is, the object may vibrate in operation while the wall could be steadfast.
3. Growth of the Objects (during operation): This could be the result of thermal expansion.
4. Overall (outside) size constraints: Internal clearance distance will be affected by the overall size. That is, with a given overall size, the distance between objects will have some particular limit. The distance between objects will be a function on the size tolerances of the objects and the tolerances on the Object locations. If the overall size is not constrained (rare instance), Object size and clearances between Objects will determine the overall size.

2.4.2 Object Arrangement

The designer usually works to minimize the overall dimensions of the enclosure by a "productive" arrangement of all of the Objects needed to fit within the enclosure. This can be done in two dimensions (X and Y) and the 3rd dimension, Z. Other arrangements of Objects look to fulfill assembly, servicing, aesthetic, or user interface needs.

In order to minimize the overall dimensions, some distance between Objects is chosen. This distance can be first thought of as a nominal distance. This nominal distance can then be adjusted to suit the design. For example, one could assume a nominal distance between objects of 0.100 inch (in all directions). Of course, the gap size would not have to be the same between all objects. Perhaps the 0.100 "gap" between objects produces an overall dimension that exceeds the expectations of the product (exceeds the product specification). Then, the designer would look to reduce the 0.100 inch gap – but, the gap cannot be less than zero, and it cannot be less than any "worst-case" problem such as an Object being supplied at the upper end of its size tolerance or other factors explored below.

Then, the designer checks to see that all of the Objects in the enclosure have been placed and that the gaps between Objects are such that all interference between Objects is avoided under all environments and user experiences that the design will exist in. The designer will also check to see that the Objects can be assembled into the enclosure in a "forthright" manner and that the service objectives of the product are upheld.

The design is ready for the Design Review Process.

Reviewing, gaps between volumes (or objects) are a function of:

- Fabrication tolerances: A given "box" may be specified as a nominal dimension. However, a slightly larger (or smaller) box results when the supplier fabricates the box to the allowable outer limits of the nominal dimension.
- Cooling requirements: A certain component may have to be spaced a minimum distance from another component so that this component is not thermally affected to an intolerable extent. In some heat-dissipative situations, components must be placed as close as possible (attached to each other).
- Assembly and serviceability requirements: Components may need certain spaces between them due to clearance required to either assemble or disassemble the components.
- Future additions to product (options): Volume may be required for planned additions or product options.

Looking back at our original intention, to locate an object, 0.100 inch from a wall, we can see that, when we get into the *detailed* design, we will have to be careful with this 0.100 inch nominal clearance, (shown by the above discussion on tolerances) as this distance can easily "shrink" (in its worst case) from 0.100 inch to 0.100 minus 0.025 (=0.075 inch). Of course, it could be *increasing* to 0.100 plus 0.025 (=0.125 inch), also. In the "sketch" design phase, we wouldn't be that concerned with this dimension; again, it would become more important as the design moves to the prototyping phase.

All of the above concentration on this 0.100 inch dimension is meant to illustrate that "some distance" is designed between objects (in this case, an object and a wall). In most designs, the overall size of the object must be minimized. This leads most designs to have the *least* possible distance between objects as possible. Examples of designs where overall size (and resulting weight) are minimized would be computer housing, coffee maker, or other household appliance. We live in a world where smaller size (usually) equates to:

- Smaller weight (better fuel savings or ease of use)
- Smaller ecological footprint (savings on materials)
- Saving of space in space-limited situations
- Lower costs (for consumer or producer)

In some cases, it will not be the *least* possible distance that is desired. Complications such as heat dissipation, or mechanical coupling (say, in a gear drive), certainly affect the distance between objects. We have been "simplifying" the design process in our examples.

So, for our example of the 0.100 inch distance between the object and the wall, the designer would actually be challenged to determine what minimal distance this could be (e.g., if this distance was 0.050 inch, our overall product could be smaller). Could this distance shrink even further? (Keep in mind that we came up with an "uncertainty" in this distance in the amount of 0.025 inch.)

In the "sketch phase" of the design, it may not be important to determine this distance *exactly*. In the interests of proving the *overall* design in a very quick manner, the designer may make this distance 0.125 inch and get into the details of reducing this distance as the design proves some success.

The wall thickness is usually a function of:

- Strength required for product operation
- Weight constraints for product
- Fabrication technique

Wall thickness doesn't have to be "constant," that is, it can vary either by the addition of ribs or gussets or the fabrication method that may allow local variation in thickness.

2.4.3 Object Arrangement Example (Fig. 2.3)

So far, in our discussion of locating two Objects (a wall and an object), we have been simplifying the discussion with just two dimensions. We will expand into three dimensions in this section. Let's take our example of Object arrangement a step further. Let's take a look at several ways to locate two objects inside of an enclosure and see what options we have. For the purposes of this example, let's say that the two Objects are both "bricks" (literally a brick), with the approximate dimensions of:

- 2.5 inches thick
- 3.5 inches wide
- 8.0 inches long

In our 2D example, we will forget about the "thickness" and just use the 3.5×8.0 width and length dimensions. So, we basically have a rectangle that is 3.5×8.0. (We will come back to the 3D example, further on, as this adds more choices for us.) See Fig. 2.3.

Now, let's say that our basic starting point in the design is to house two bricks, Brick A and Brick B (both with the same dimensions), in an enclosure. At this initial point, we have no constraints of:

- Overall size or shape of enclosure
- Cost of enclosure material

Optimal Object Placement

Examples of different arrangements of two objects (Brick A and Brick B) in an enclosure

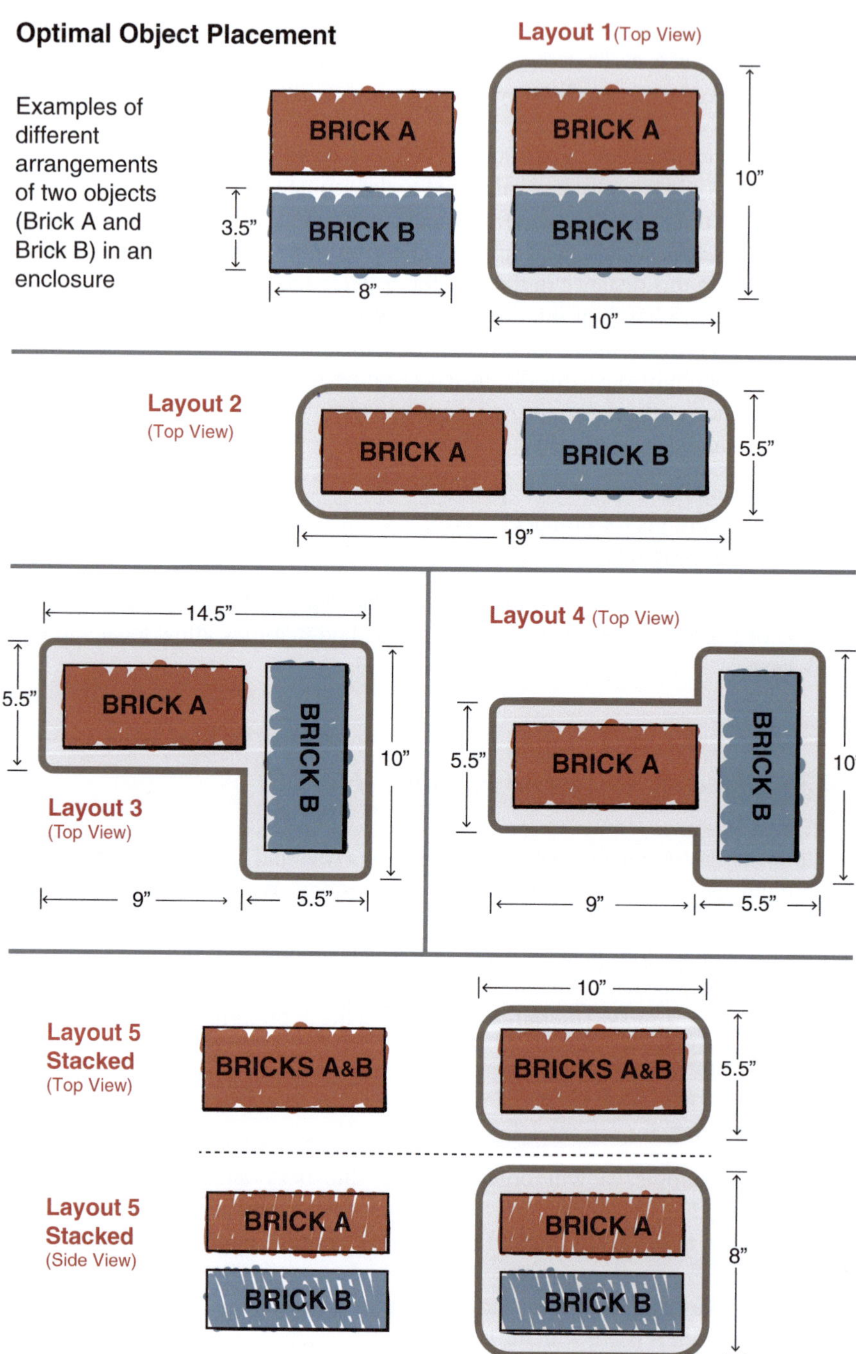

Fig. 2.3 Optimal object placement

We can easily envision (at least) *five* different ways to locate Brick A and Brick B relative to each other, producing very different enclosures. Of course, there are more than five different ways, but I've chosen the standard "Cartesian" arrangement of bricks in which they are parallel or aligned to each other. Let's look at these *five* different layouts and comment a bit about why one might have some advantages (over the other layouts). A constant *one inch* is assumed between Brick A and Brick B, and that same *one inch* clearance is assumed between a brick and a wall (side, top, or bottom).

- Layout 1: Brick A and Brick B side by side along width
- Layout 2: Brick A and Brick B aligned along length
- Layout 3: Brick A and Brick B in "L shape"
- Layout 4: Brick A and Brick B in "T shape"
- Layout 5: Brick A and Brick B "stacked" (a 3D version – this is the only layout that has used the "3rd Dimension")

Now, let's analyze the five layouts. (For all of the layouts, let's assume a very thin-skin for the enclosure that adds essentially zero to the enclosure width, length, and height. Also, we'll neglect the rounded corners that an enclosure might have, and just assume square corners).

- Layout 1 is seemingly the simplest layout, ending with a relatively square shape for an enclosure. Resulting enclosure is $10 \times 10 \times 4.5$ high. The six sides (areas) are $2 \times (10 \times 10) + 4 \times (10 \times 3.5)$.
- Layout 2 is "long," rather than "square." This type of enclosure may have unique applications, such as for better utilization of desk space. Resulting enclosure is $19 \times 5.5 \times 4.5$ high. The six sides (areas) are $2 \times (19 \times 5.5) + 2 \times (19 \times 3.5) + 2 \times (5.5 \times 3.5)$.
- Layout 3 may be better suited for "corner" applications. Resulting enclosure is $(5.5 \times 10 \times 4.5) + (5.5 \times 9 \times 4.5)$. The eight sides (areas) are $3.5 \times (4.5 + 9 + 10 + 4.5 + 5.5 + 14.5) + (9 \times 5.5 \times 2) + (10 \times 5.5 \times 2)$.
- Layout 4 is similar to Layout 3 but with a more symmetrical look (same Volume and same perimeter as Layout 3).
- Layout 5 offers minimal "floor plan" but results in the tallest of the layouts. Resulting enclosure is $10 \times 5.5 \times 8$ high. The six sides (areas) are $2 \times (10 \times 5.5) + 2 \times (10 \times 8) + 2 \times (5.5 \times 8)$.

These simple layouts illustrate that even the placement of two Objects, in this case, two bricks, represents quite a few possibilities. If one would add a 3rd Object or Objects of different sizes, one can see that this gets quite complicated. Sometimes, there is a fundamental reason for the placement of one object relative to another, as one object's "in" should be close to the other object's "out" (nesting of objects). In any case, let's continue to discuss some relative merits of the five layouts, above.

There may be a fundamental aesthetic rationale for a choice between the layouts. That is, Layout 1 could be chosen because it is deemed "more direct" (more "honest"), and it is possible that Layout 3 is deemed more "interesting." So, the choice of layout could come down to a marketing decision that the customer will find a particular enclosure shape to be more pleasing (and, thus, results in higher sales of the product).

How about optimization (minimization) of the surface area of the enclosure – which layouts results in the minimum surface area? Again, the enclosure surrounds both bricks on their sides, top, and bottom.

	Volume	Perimeter area
	(Cubic inch)	(Total square inches)
Layout 1	450	340
Layout 2	470.25	380.5
Layout 3	470.25	377
Layout 4	470.25	377
Layout 5	440	358

Some Conclusions to Object Arrangement

If weight is to be minimized, Layout 1 is best as the weight is primarily due to enclosure perimeter area. (Bricks have the same weight in all layouts, and the air is negligible.)

If volume is to be minimized, Layout 5 is best. This may be advantageous for some designs where size is restricted.

Volume is the same for Layouts 2, 3, and 4.

Largest perimeter area is Layout 2. An iteresting "Layout 6" (not shown) would be a enclosure that would be a sphere. A 6 inch radius sphere could house Layout 5 with the resulting volume of about 900 cubic inches while only having a perimeter area of about 450 square inches - perhaps a "creative" solution?

Obviously, as more "Objects" are added, it will get more difficult to optimize the objects for volume and perimeter area. The designer's ingenuity comes with "nesting" various geometrical objects in a given layout area.

Various techniques can be used to optimize the layout's compactness. Once all objects within the enclosure are determined, the designer can model those objects and start placing them in an orientation that:

1. Utilizes the space in an efficient manner.
2. Places objects that require nearness to each other, as closely as possible to each other. This may be due to mechanical, thermal, or electrical reasons to do this. For example, if a cable connects two Objects, it may be advantageous to have these two Objects as close as possible – how about eliminating the cable by connecting (directly) Object 1 into Object 2?
3. Places objects that need as much distance from each other, as far away as possible from each other. Again, there may be mechanical, thermal, or electrical reasons to do this.

It should be noted that Layout 3 and Layout 4 could be more complicated to produce (manufacture). The nonsymmetrical enclosure could be more difficult to fabricate than straight walls. This is not necessarily so if the enclosure is a tooled (molded or cast) product. However, if the enclosure is sheet metal, the extra walls provide a bit more of an issue to fabricate.

A note about designing in 3D: As all of the objects we design are actually 3D objects, we will actually need to design in 3D. A quick sketch in 2D may solve a portion of the design intent, but all of the details will need to be shown as the design moves to 3D. This makes working in CAD essential to the accuracies and speed required in the contemporary design world. Creating 3D models of all of the objects makes checking clearances and fits simple to change and optimize. An example of this is how a mechanical designer views a printed circuit board assembly (PCBA). It basically has a 2D layout of many components mounted on a printed circuit board. However, all of those components are at various heights, thus clearances above and below the PCBA will vary in many areas. Thus, the PCBA, while primarily thought of as a "2D area" (in the "sketch phase"), has a "thickness" to it that makes it a 3D volume.

Chapter Summary

This chapter takes us from the beginning point of a design where we have just an idea. It shows us how we transform that idea into the geometric placement of objects into space that gives us a physical manifestation of that idea.

We started by looking at our starting point and defining the boundaries of the design – what do we start with and what is the "outer edge" of the design. We have to define what the product is that the customer will need.

We saw how designs proceed from Revision 1 to Revision X, where X is the design that provides what we believe to be what the customer will need.

Finally, we took a look at seeing how the individual objects that need to be in the design can be optimally placed to solve customer needs. There are trade-offs to be considered, and we must be aware of how we determine the best choices between these trade-offs.

Chapter 3
Structural Considerations

In the previous chapters, we defined what a successful design is and then moved on to determining the placement of the objects that will be in the design. We'll now take up the structural considerations of the design. Why consider the structural considerations at this juncture, and why not the thermal aspects, or the user interfaces? It's probably only because I have a mechanical engineering background so I "naturally" first see how the design must be "structurally sound." I feel we must build upon a "solid foundation," so that the rest of the design can build upon that. The electronic enclosure (itself) is, of course, a structure that must be strong enough to work in the various environments that the customer (user) will be using the product in. So, let's begin with a discussion of the main considerations of providing this "solid foundation." This chapter will focus on:

- Using strength of material concepts to propose structural solutions
- Defining a generic process for considering the structural design of our electronic enclosure
- Look at some examples that specifically illustrate the general concepts We'll close this chapter with a section titled "Bonus Section". This last section is meant to add some complications to our problems on strength of materials and also to show how other considerations besides strength will be important to our design choices.

3.1 Introduction: Strength of Materials

This chapter is not an attempt to review all of the principles of strength of materials or of mechanical engineering. Entire texts have been devoted to stress, strain, and strength alone; therefore, we will just "scratch the surface" of knowledge and emphasize how some basic equations help our job to design electronic enclosures. However,

© Springer International Publishing AG, part of Springer Nature 2019
T. Serksnis, *Designing Electronic Product Enclosures*,
https://doi.org/10.1007/978-3-319-69395-8_3

in the course of design of these electronic enclosures, an exacting knowledge of the structural considerations will be essential to the overall success of the design.

The reader does not need a mechanical engineering degree or be an expert in strength of materials to benefit from this chapter. I'm hoping that some of the basic principles are touched-upon, enough to give the EPE Designers some value no matter where they are in their career. It was my belief that the more that designer understands basic strength of materials, the better the enclosure design would be.

For example, the EPE Designer can design an enclosure using 1/8-inch-thick aluminum for the enclosure material. Testing may prove that 1/8-inch-thick aluminum does indeed pass the shock and vibration testing. However, here are some questions to ask about this choice of thickness and material for the design:

- Could we have used 1/16-inch-thick aluminum instead, which would have saved weight and possibly been easier to fabricate?
- Could we have used 1/8-inch-thick plastic instead, which would again have saved weight and possibly been easier to fabricate?

So, as you can see from the above questions, it is not enough to just solve the problem, we need to solve the problem in the most *cost-efficient* manner possible. We'll get into the detail of "the most cost-efficient manner" at the beginning of Chap. 4. But for now, we'll concentrate on determining suitable designs that are at least structurally successful.

One of the biggest contributions that a designer will make to the design of the electronic enclosure is data to prove that the design will hold up "structurally" to the rigors of the customer product environment. I'm hopeful that whatever the reader's background is, they will be able to propose a design for an electronic enclosure that will be strong enough to pass the rigors of testing. I'll introduce some of the basic equations and concepts involved to help even the beginning EPE Designer, and hopefully it will also help the veteran EPE Designer.

The fundamental approaches for designing a suitable structure for an electronic enclosure break down into *four* basic approaches:

1. Take a look at similar products that already exist, and use the solution already designed as a quick starting point for the design at hand. Pluses for this kind of approach are speed, but the downside is that your design may suffer due to lack of creativity toward solving a unique problem that your specific product should solve.
2. Quick, "back-of-the-envelope" design. This approach uses some rudimentary design equations on simplified structural elements. We'll explore some examples of these design approaches with some example problems later on in this chapter.
3. More complex analysis. This is explored a bit more in Sect. 3.3 on "Analysis Required." Again, this text will not cover much ground for designs requiring complex analysis. What I would like to emphasize in this chapter is a feel for the structural elements of the design and what some "quick fixes" would be for improved designs.

4. Overdesign – Of course, *overdesign* is not the correct answer for all of the designs. I've already touched upon this above in the example on the solution using 1/8 inch aluminum. I'll go into another example of overdesign below. In a *competitive* product market, where customers make buying decisions mainly on price, overdesign will likely lead to increased product cost (or, certainly, increased weight and size). Structural overdesign is basically starting with a design that has a very good likelihood of success of passing structural testing, that is, surviving the customer usage environment for shock and vibration.

A lot can be said for *overdesign*. The EPE Designer could determine that a bracket that is 18 gauge (0.048) thick metal would "do the job" but, *instead*, choose 16 gauge (0.060) thick metal. Increasing the thickness of the bracket gives one some comfort for a couple of reasons:

- That the design will stand up to some of the forces that are *not known* to a high degree of precision. This will be further explored in Sect. 3.2 on "Design Process."
- That there is just a "factor of safety" greater than 1.0 in the design. A factor of safety *equal* to 1.0 means that your design *just* meets the design criteria. A discussion on the considerations of designing with increased factor of safety is covered in Sect. 3.2 of this chapter.

Also, it is possible that there may be some economical reason for placing 0.060 thick metals in the design. For example, if the majority of the design is 0.060 thick already, and if the bracket can be made out of a piece of "scrap," a savings might result.

It is very possible (in the above example) that using 0.048 thick metals *with the addition* of some simple "ribs" or bends would make the design *much stronger* than 0.060 thick metals. This is what I would like to spend some time on, and this issue of adding ribs to a design is explored in the problem shown in Sect. 3.4.2.

3.2 Design Process for Structures

I'd like to give the reader a generic process for designing the electronic enclosure (or, an *individual part* in the enclosure) that will satisfy the structural considerations of the design. By going through these six steps, the designer should be ready to propose a material and cross section that will work. I'll individually break out the six steps as subsections.

3.2.1 Similar Designs

How have other designs in the industry handled similar situations? The other designs could be from examples within your own company (past products) or from competitive products outside of your own company.

3.2.2   Forces on Part

Determine the forces (static and dynamic) on the object – amplitude and direction of those forces. The part's own weight generally doesn't come into consideration in electronic enclosures for static forces but does get considered for dynamic forces. In this text, I refer to "objects," "parts," and "members," but they should all be considered being one-in-the-same.

3.2.3   Existing End Conditions

Determine the "end conditions" of the object, that is, its degrees of freedom of movement, and how the member will be supported. Common end conditions are "fixed" (not allowed to move) or "free" (allowed to rotate). End conditions have an effect on determining the amount of stress that loads will create.

3.2.4   Propose Material and Cross Section

Determine the material and cross-sectional combination needed to support those forces (from Sect. 3.2.2), keeping in mind that "strength" is an inherent aspect that belongs to materials (so, the higher the yield strength of the material, the more load-bearing ability that material contains), and forces produce stresses in those materials. All materials have limits for maximum stress where we either have the start of deformation (yield strength) or complete failure point (ultimate strength).

Maximum stress in the member is generally known by the "common equation" of

$$\sigma = Mc \, / \, I$$

where:

σ is the maximum stress in the member.
c is the distance of extreme "fiber" from bending axis.
I is the moment of inertia. This is a property of the cross-sectional area of the object.
M is the maximum moment in the cross section furthest from where the force is applied. It is that force times its distance from an end-point condition to where the force is applied.

Basically, only *two* choices initially exist to design higher load-bearing members (as the terms "c/I" are both related to cross-sectional area and that area's "dispersal" away from the "neutral plane" of bending).

- Change the material, which allows a change to the stress limits. So, choosing a material with higher stress limits allows more loading to be placed on that member.

- Change the material's cross-sectional property, basically the member's second moment of area (also known as the moment of inertia, I) and the amount of area that can be concentrated away from the member's "neutral axis" or centroid. Increasing area will essentially increase a member's ability to carry more loads. Increasing that area away from the member's "neutral axis" will also help the member carry more load (which is why "I-beams," which have a lot of the member's cross-sectional area very far from the "neutral axis," are excellent load-carrying members).

I've illustrated the interrelation between changing both material and cross section in Fig. 3.1. Here, we have a very common load situation, one where a force is acting on the end of the member, and the member has a fixed end condition. We will be showing how various changes in (either or both) material and cross section can solve a problem. The basic problem is finding a member strong enough to survive this load, a 2000 pound force. The EPE Designer is tasked with determining both the material and cross section of the member so that the maximum stress *in* the member will be under the maximum stress (let's say yield stress) *allowed* by the particular material. So, we can utilize the equation from above as a starting point in the design:

$$\sigma = Mc / I$$

We can calculate the maximum moment, M, as being equal to 48 inch × 2000 pounds, which we'll then say is 96,000 in-lb (this will be the same value for any material and cross section that we choose). Let's put forth two candidate materials:

Pine wood, which has a yield stress of 1200 pounds/in^3 (psi)
Aluminum, CR H-18, which has a yield stress of 22,000 psi

Let's keep to simple rectangular shapes, which have the moment of inertia value of (for either material):

$$I = bh^3 / 12$$

where
  b is the width of the member and h is the height of the member in cross section.
  In this example, c, which is the distance from the extreme fiber to the bending axis, will be $h/2$.
  Thus, our equation for stress becomes:

$$\sigma = Mc / I = \left(96000 \times \left(h / 2\right)\right) / bh^3 / 12 = 576,000 / bh^2$$

(Note that the stress in this member is dependent on the height of the member *squared*, which underscores the need for high "aspect ratio" (the ratio of height to width) cross sections.)

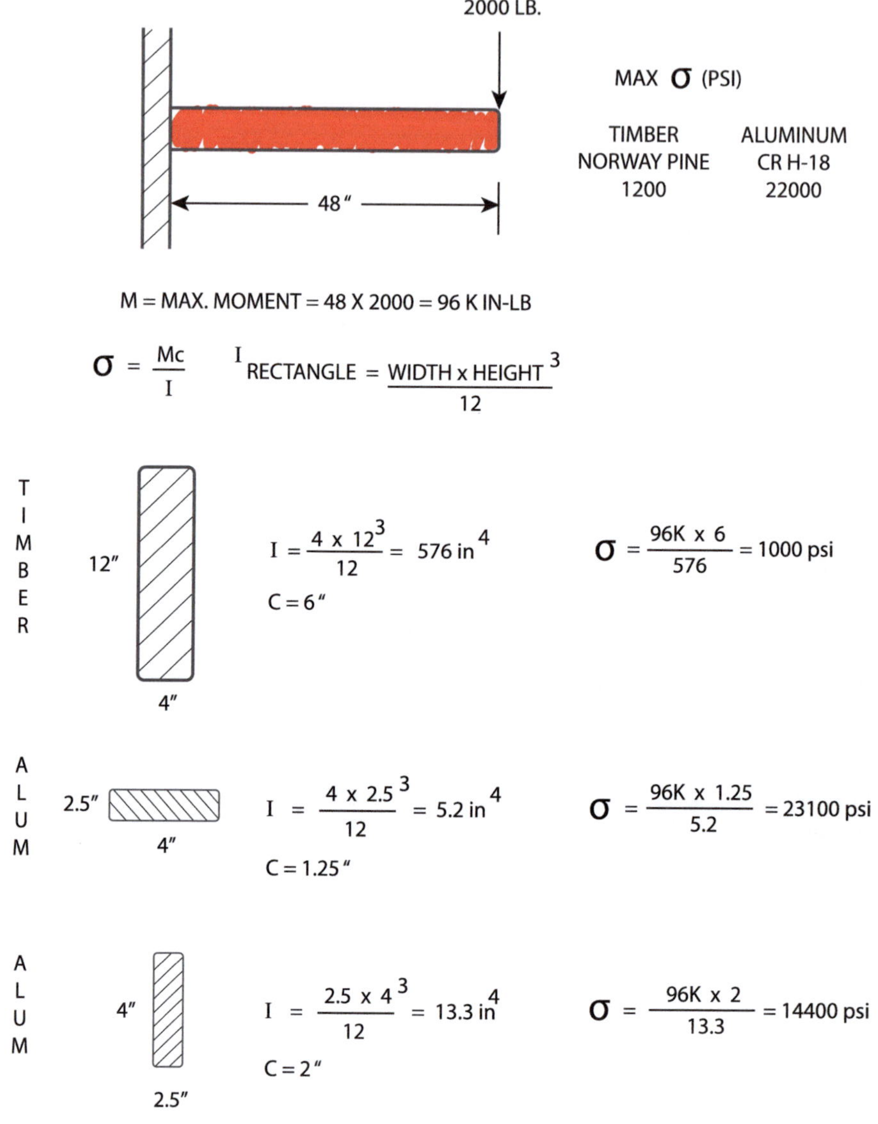

$$\sigma = \frac{Mc}{I} \qquad I_{RECTANGLE} = \frac{WIDTH \times HEIGHT^3}{12}$$

$$I = \frac{4 \times 12^3}{12} = 576 \ in^4 \qquad \sigma = \frac{96K \times 6}{576} = 1000 \ psi$$

$$C = 6"$$

$$I = \frac{4 \times 2.5^3}{12} = 5.2 \ in^4 \qquad \sigma = \frac{96K \times 1.25}{5.2} = 23100 \ psi$$

$$C = 1.25"$$

$$I = \frac{2.5 \times 4^3}{12} = 13.3 \ in^4 \qquad \sigma = \frac{96K \times 2}{13.3} = 14400 \ psi$$

$$C = 2"$$

Fig. 3.1 Material & cross-section choices

3.2.4.1   Pine Wood Solution

Let's try to design the member that will be made of pine wood. By putting in:

$b = 4$ inch and $h = 12$ inch, we see that the maximum stress will be 1000 psi. This member (pine wood, with a cross section of 4×12) has a "stress limit" of 1200 psi, and the load on it is only 1000 psi. Nice, we have "overdesigned" the member (with a factor of safety of 120%).

Now, the EPE Designer needs to look at "other" design constraints (like weight or cost) to make a decision to see if this pine wood beam will be a good candidate for our electronic enclosure.

Spoiler alert: We'll discuss the 15 considerations for determining material selection for any part in Chap. 4, but for now, just look at weight as another consideration for the "final choice" of material and cross section.

Let's look at the weight of this pine wood beam. At 30 pounds/ft^3, the beam will be 40 pounds. Fine (for now).

3.2.4.2 Aluminum Solution

Design is all about presenting some logical choices, so let's look at an aluminum beam.

We can choose,

$b = 4$ inch and $h = 2.5$ inch. We can see that the maximum stress will be 23,100 psi. This is *above* the maximum yield stress for the aluminum, so *this will not be* structurally *satisfactory* in our design.

But what about remembering that the height of the beam is the larger "factor" in our calculations for moment of inertia,

$b = 2.5$ inch and $h = 4$ inch? This will be the *same* cross-sectional area as the previous example for the aluminum beam. Now, the maximum stress will be 14,400 psi, well within the maximum of 22,000 psi for this aluminum. Thus, "rotating" the same cross section, where the thicker aspect is in the direction of the load force, allowed this choice of material and cross section to be structurally successful.

Let's look at the weight of this aluminum beam. At 169 pounds/ft^3, the beam will be 47 pounds. This compares to 40 pounds for the pine wood.

In summary, we have looked at how two different materials (pine wood and aluminum) could be used to solve the structural problem. We can develop cross-sectional areas for each material that solves the structural problem.

In design, deformation often shares an equal importance with strength. A load member may have sufficient strength to withstand a particular load, but it may deflect an unacceptable amount beyond the elasticity of the engineering material. Problems, where deflection (and thus the material's modulus of elasticity, E) is also under consideration, are shown in some examples further on in this chapter.

The economics of the above choices (change material or change material cross section) pose an interesting problem to EPE Designers. Many combinations of material and cross-sectional area will work, but a choice must be made that fits the overall goals of the project. Besides functioning, it must meet project goals of cost, manufacturability, risk, weight, time to market, etc. These choices will be further investigated at the beginning of Chap. 4. It is possible that alternative solutions would need to be reviewed, tested, and prototyped. One of the biggest assets a designer can bring to the design would be to quickly find the logical choices to be made among the viable candidates for material/cross-sectional choice that will solve the problem at hand.

3.2.5   *Combine Function*

Can the part being designed be *combined* with another part in the assembly which is adjacent to this part? Basically, can two *separate* parts (being envisioned) be combined into a *single* part? This is illustrated in Fig. 3.2.

The "alternative thinking" aspect of looking at the part being *combined* is to actually look to create two *separate* parts from a (envisioned) single part. This could result in a lower overall cost reduced solution to the *combined* design.

One of the main choices (for a candidate material/cross-section solution) will be determining how to fabricate this solution in production. For example, some of the choices involved here are:

- What is the tooling budget for the project? Can the project "afford" spending an amount of capital needed for casting, injection molding, extruding, or other fabrication techniques that may be under consideration? Is there existing tooling that can be utilized? A determination must be made to find the "payback period" of a tooled solution. For example, knowing:

  1. How much tooling will cost
  2. How many parts will be needed (over the product "lifetime")
  3. How much un-tooled part will cost
  4. How much the tooled part will cost

will determine when the "payback period" of the tooled solution. For example, if tooling will cost $50,000, and the un-tooled part cost is $10, while the tooled part cost is $1, this would result in a savings of $9 per part needed. Thus, the tooled part pays for the tooling in 50,000/9 = approx. 5500 parts. If 5500 parts are expected to be sold in a year, then the "payback period" will be approx. 1 year. See previous discussion on tooling "breakeven" in Sect. 1.7.

- Are there "off-the-shelf" or "previously designed" solutions that could be used in the design? This could save the tooling cost, and the increased volumes of this "new" usage (when combined with the "old" usage) will lower the individual piece cost.

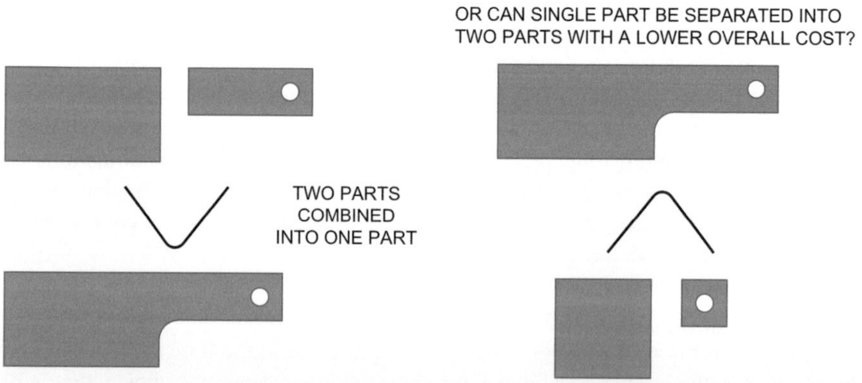

Fig. 3.2 Combining (or separating) parts to improve design

- Can the fabrication technology be phased-in over the course of the project? That is, can we use one fabrication technique in the prototype/first production stage (e.g., CNC milling) and then switch over to a tooled solution (e.g., casting) after first production? With this scenario, cost reductions are phased-in, and the savings doesn't occur near term (but, rather, in a longer time period).

3.2.6 Determine Factor of Safety Needed

A determination of "factor of safety" must be reviewed at this time. That is, answers to the following questions must be known:

- If the part fails, does anyone get injured? What is the cost of an unpredictable failure in lives, in dollars, and in time?
- How critical is this particular part in the overall function of the product? If this part fails, does the entire product fail?
- How well are the forces known (from Sect. 3.2.2 above)? Do we know the "error bars," that is, how much the forces can deviate from the assumed nominal value?
- Determine the "critical aspects" of the chosen design (material or geometry), and how in-production will they be specified, certified, and inspected? Make notes to assure these steps (certification/inspection) will be done. Determine the testing required in the various stages of the design that will be required to assure that the final design will be adequate for shipment to the customer in production.
- There will be an optimized solution which generally can be found by analyzing the major components of the design and determining where the "weak links" in the design exist. This can be found by utilizing some testing methodologies that induce failures by testing beyond the environmental limits (such as highly accelerated life testing, HALT). By first identifying where failures might occur, then by testing design prototypes, data can be generated to determine whether certain segments are near their design limits.

If any of the above six steps in the design process do not have answers known to some degree of confidence, the designer is faced with:

- Making further inquiries to get better information.
- Going forward with the design. It would be rare for designers to know about all of the forces and interrelation of parts at the very beginning of the design process. Certainly, the designer can list the assumptions made and the additional information that would be essential. It is certainly possible to design parts, prototype the parts, and test them under the conditions that they will need to function in. Several approaches to this dilemma of "going forward with the design without knowing all of the information" can be taken; let's explore an example where:

Design 1 has a weight that is 110% of target weight but has a 95% chance of being structurally successful. Design 2 is 100% of target weight but has a 75%

chance of being structurally successful. So Design 1 is 10% over the target weight, but with a much lower risk of failing to meet the design goal of working from a structural point of view.

So, what is being "traded-off" is the time needed to optimize the design. Certainly, the product must work from a structural basis. It will be difficult to determine the "margin" in the design at the very beginning of the program. Going forward with the design *without knowing all of the information* has value in that the "basic design" can be tested. It is hoped that the "basic design" can be modified in a quick time frame that allows the program to continue as the rest of the information is attained. We can move forward quickly by "overdesigning" the parts or invest more time to "marginally" meet all of the requirements. These two paths are investigated a bit more below:

A. "Overdesign" the parts – this approach probably guarantees that the parts *will structurally function* under testing. The idea here would be to iterate back to a *less* conservative design as testing reveals where material and weight savings are appropriate. This approach at least maximizes the chances of the design meeting the structural functionality requirements very early in the test phases of the project. However, weight changes to the design to bring these parts closer to "marginal" structural success will require time (and money) to retest the design to validate the changes. Most projects have limited time for iterative approaches to attain parts that are "perfectly" designed.

B. Design parts with the more time-consuming path of "just marginally" meeting both the weight and strength requirements. So, this stratagem is different than overdesign (above) in that the parts are designed that have a chance of (just barely) working. For example, if space and weight reduction are highest on the list of product requirements, a design that is "marginally" acceptable from a structural strength factor, but has a greater material and weight savings, may be what is needed. This approach attempts to balance both "risk and reward" and should have the agreement of the design team to go forward. With this design, the material and weight goal would be met. However, risk of this design *not structurally working* goes from 5% to 25%. So, the "B" design path shows higher risk of not meeting the product requirements for structural strength but will meet the product requirements for weight.

C. Blends of the above two approaches may be appropriate. That is, some parts of the design would be conservative, while other parts of the design would be more risky. This perhaps allows an "overall risk tolerance" to be a part of the overall design. Experienced design teams will know the best places in the design to "push the envelope" of acceptability.

3.3 Analysis Required

There are certainly many designs that warrant the most exacting analysis in the design of electronic packaging. In any highly competitive product design area, it will be the company that does the most productive job with a given technology that will maximize its chances for success. The very highest degree of analysis will be needed if the product has:

- A "high" production quantity. If hundreds of thousands of a particular unit are to be produced, then the savings of a dollar per unit could result in substantial total savings. An analysis that saves even a small amount of cost will result in a lot of overall profit due to the larger production quantities. If, however, only a few units are to be produced, the potential for savings is greatly reduced, and, once a design is deemed to be functional, a large investment in cost reduction will not bring substantial savings.
- A high degree of safety as a requirement due to the environment that the product will be placed into. Examples of this are products that are in the transportation, utilities, medical, or educational industries. All customers need to have a safely operating product.
- A "mission" that is critical to the customer. This would include products needed for military, space agency, or government in general.

Note here that there can be no excuse for a design that is so overdesigned that it lowers the profitability of the company. Designers and engineers should be ever vigilant to the possibility of cost reduction. The elimination of parts, the design for manufacturability, and the overall elegance of design lead to product leadership. It is in the first stages of design that present the *most* cost reduction possibilities. As the design progresses to even the prototype stages, the cost of redesigning for cost reduction starts to rise exponentially. More on this aspect will be presented in Chap. 6 on "Assembly and Serviceability."

Also, a note on safety is appropriate. There can be no excuse for underdesigning a product in any area where safety is an issue. Underwriters Laboratories (UL) and other safety agencies, of course, certify electronic equipment for safety consider-ations. That is, a safety agency will take a product (specifications and working units) and subject them to both review and testing. Most electronic products, certainly those sold worldwide, will have to pass rigorous agency approval certification. More on this aspect will be presented in Chap. 10, "Safety by Design."

The number one design consideration is still and will always be *functional-ity*. That is, the part must *function* as it is intended. It doesn't matter how well it looks or how elegantly it can be produced, IF the part will fail under load. This is a major reason why the loads must be understood by the designer.

Modern analysis software solutions using finite element analysis (FEA) are very ubiquitous. A search on Google reveals introductory material such as:

A. *Finite Element Analysis*, by David Roylance, MIT. Describes the three principal steps as:

  - Preprocessing, where a model of the part to be analyzed in which the geom-etry is divided into a number of discrete subregions, or "elements," connected at discrete points called "nodes"
  - Analysis, where the dataset prepared by the preprocessor is used as input to the system of linear or nonlinear algebraic equations that calculate the stresses and displacements

- Postprocessing, where the results are graphically displayed to assist in visualizing the results

B. *Linear Analysis*, by Professor K. J. Bathe, from the MIT *open courseware*, MIT. This video series is a comprehensive course of study that presents effective finite element procedures for the linear analysis of solids and structures.
C. *Finite Element Analysis*, Dr H. J. Qi. Describes the FEA process as:

- Formulating the physical model, that is, describing (perhaps, simplifying) a real engineering problem into a problem that can be solved by FEA
- Using the FEA model by discretizing the solid, defining material properties, and applying boundary conditions
- Choosing proper approximate functions, formulate linear equations, and solving these equations
- Obtaining results in both numerical and visual formats

There is no doubt that using FEA can provide much useful information about engineering problems involving structural analysis (along with solid mechanics, dynamics, and thermal analysis). Any answers coming out of this analysis should be first tested by using simplified models and forces to see if the answers make some sense. Testing should be used to verify the assumptions made and the resulting answers. Another attribute of using FEA analysis is that small changes in the design can also be inputted into the analysis to see how the results vary. In this manner, it can be shown very quickly how to make the design better.

Some companies are large enough to have an entire department devoted to FEA analysis, while others operate with the expectation that the designer will be analyzing the structures using FEA on their own.

3.4 Structural Problems: Static Loads

Again, as this text is not meant to cover all of the various structural considerations or problems encountered, I'd like to highlight a few problems (I kept it to three) that highlight the following:

- Problems that can be thought of as either "individual" in that loads and forces are applied to *single members* being designed or the "structures as a whole," that is, it could be an analysis of the entire (assembled) structure.
- Many suppliers of individual parts (or sub-assemblies) offer design guidance in their own literature that is certainly available to the individual designer. Much of this information is based on both empirical and analytical experience gained over the years. The designer is cautioned to understand the background and limitations on any of this information. Some of the information that is presented in graphical or tabularized formats is rooted in fundamentals, but this may not be apparent. Some examples of this type of available "supplier data" is in the design of plastics, seals, EMI components, and bearings (to name a few).

Static loads on members in electronic enclosures are due to:

1. The member's own weight
2. Loads applied by other members
3. Loads due to thermal effects, residual stresses, etc.

Static loads will induce members to fail either by applying a force resulting in:

1. The yield strength of the material being exceeded.
2. An over-deflection of the member which results in the member performing outside of the design intent. As all loads produce some deflection, it must be known at some point in the design just how much deflection by the member is to be allowed.

Failure by fracture under static loading is not as common in ductile materials as in brittle materials. In a ductile member, failure usually occurs as a result of excessive inelastic action which leads to very large overall deformations long before fracture.

Dynamic loading will be covered in Sect. 3.5. Dynamic loads are generally those that vary with time, whereas static loads do not change significantly in a relatively short time period. Some dynamic loads common in the design of electronic enclosures are repeated loads, impact loads, and energy loads. Energy loads are loads that are expressed more easily in terms of the energy transmitted during the impact period (than in terms of applied force).

Fracture caused by a repeated load is commonly referred to as a "fatigue" failure. Vibration can be a cause of a fatigue failure.

Topics generally covered by the following three problems are:

1. What is a beam (vs. a plate)?
2. Stress formulae and maximum stress.
3. Deflection formulae and maximum deflection.
4. Section modulus.
5. Modulus of elasticity.
6. End conditions.
7. Load conditions.
8. Worst-case loading.
9. Combined loading.

3.4.1   Cantilever Beam Analysis (from Tecknit EMI Shielding Products Manual)

A majority of electronic enclosure stress analysis can be characterized by calculations of a "simple" beam. But, first of all, let us define a beam. Roark and Young (see Ref. [1]) make the following assumption for the application of beam flexure formulae:

• The beam must be long in proportion to its depth, the span/depth ratio being 8 or more for metal beams, and 15 or more for beams with relatively thin webs. This

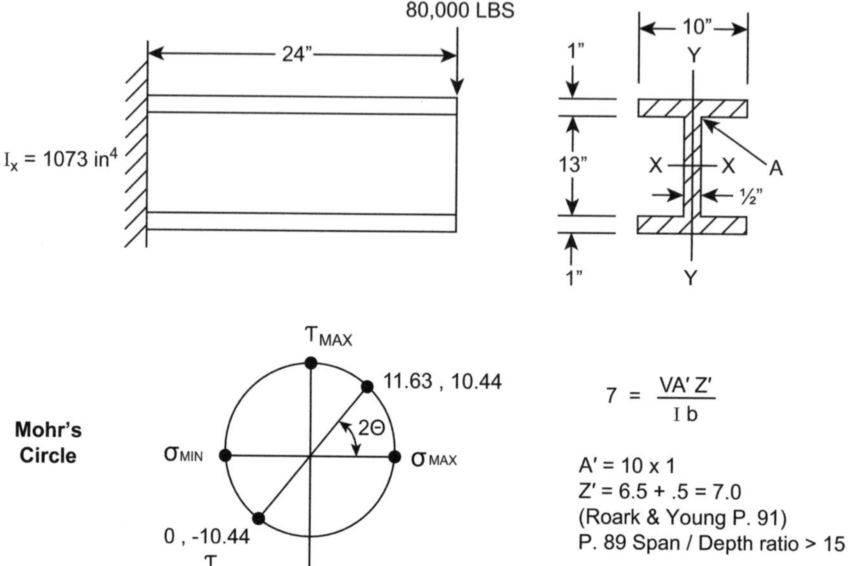

Fig. 3.3 Cantilever beam analysis

is shown in an excellent example by Byars and Snyder (see Ref. [2]), a cantilever beam as shown in Fig. 3.3. Determine the principal normal and shear stresses at Point A just below the flange. Assume elastic behavior and neglect any stress concentration at the wall.

Solution: The bending moment at the left-hand end of the beam is:

$$M = (80,000)(2)(12) = 1,920,000 \, \text{in} - \text{lb}$$

and the vertical shear force is:

$$V = 80,000 \, \text{lb}$$

Hence, the flexure stress at Point A is:

$$\sigma = My / I = (1,920,000)(6.5) / 1073 = 11,630 \, \text{psi} \, (\text{tension})$$

and the transverse shear stress at Point A is:

$$T = VQ / It = (80,000)((10)(1)(7)) / (1073)(0.5) = 10,440 \, \text{psi}$$

Mohr's circle for this state of stress gives:

$$\sigma \, \text{max} = 17,750 \, \text{psi} \, (\text{tensile})$$

$$\sigma \min = 6{,}130\,\text{psi}\,(\text{compression})$$

$$T \max = 11{,}940\,\text{psi}$$

(Mohr's Circle is used to find the total stress when bending stress and shear stress are considered in the problem) Note, however, that in such a short member as this with its thin-web cross section (span/depth ratio = 24/15 = 1.7), the validity of the flexure formula is questionable. Notice, for example, that the shear and normal stresses are of the same order of magnitude. Also, note that the length of this beam would have to be on the order of 19 feet, for the correct span/depth ratio to apply.

The importance of the above example is the emphasis on the effect of transverse shear stresses on maximum stress. In determining maximum stresses in beams, don't be satisfied with your results until you have exhausted all possible combinations of flexure and shear stresses which could give the maximum principal stresses. Often, the construction of the shear and moment diagrams and a comparison of the orders of magnitude of the flexure stresses and transverse shear stresses will greatly simplify the problems.

Using some of the beam stress formulae from the example above, we will continue with the "main thrust" of a problem that an electronic enclosure designer may face. That is determining the (maximum) fastener distance ("C") for a "cover plate" onto a housing chassis. This type of problem involves an environmental seal along the housing that will provide (see Ref. 6)

- Protection from dust, moisture, and vapors
- Adequate EMI shielding

We'll take up the shielding portion of the problem in Chap. 9. Right now, we'll tackle the "structural problem" of designing the basic seal design geometry in regard to maintaining adequate strength to provide a moisture seal. I'll be quoting some material from the Tecknit EMI Shielding Products Manual (see Ref. [3]). Note here that I am using some reference material from a "manual." This remains, even in the "Google Search" age, a very valuable source of information for designers. Many of these manuals are hardbound and were available from original equipment manufacturers for designing in their particular components. Now, a lot of this "design guide" information is available online (as opposed to being available in a hardbound manual). Usually, sales personnel of the component manufacturer are aware of the various "guides" and online information available to designers today.

Now, returning to the structural considerations of this enviromental seal problem:

A. Material of seal: covered later in Chap. 7, "Product Environments (Sealing)"
B. Cover and chassis material: modulus of elasticity covered here (corrosion covered later in Chap. 4, "Materials and Processes." Surface finishes are covered in Chap. 7, "Product Environments (Sealing)"
C. Cross-sectional area (moment of inertia needed), covered here
D. Bolt spacing, covered here
E. Compression stop, covered here

The "fastener distance" problem is solved (approximately) here in the Tecknit manual by the use of an equation (where C is the spacing between bolts).

With the three assumptions of:

1. Gasket width = cover plate width.
2. Maximum pressure (exerted by gasket) equals three times the minimum pressure (exerted by gasket).
3. Minimum pressure is 20 psi.

Comparing an enclosure made of aluminum (vs. made of steel):

$$C = 59.6\left(t^3 \Delta H\right)^{\frac{1}{4}}$$

For aluminum plate (E = 1 × 10⁷ psi).

For ΔH = 0.01 inch, a reasonable gasket deflection, and t = 0.125 inch, $C = 4.0$ inch

$$C = 78.5\left(t^3\,\Delta H\right)^{\frac{1}{4}}$$

For steel plate (E = 3 × 10⁷ psi).

For ΔH = 0.01 inch, a reasonable gasket deflection, and t = 0.125 inch, $C = 5.2$ inch.

A few further observations about the equations (and answers) are:

1. We see that the bolt spacing for a steel enclosure is more than an aluminum enclosure – this makes sense that the stiffer material would allow less flexure. A bolt is needed every 4 inches for an aluminum plate, while if we use steel for the plate material, a bolt will be needed every 5.2 inch.
2. We see that the bolt spacing varies as the cube of the thickness – we would expect that the equation (for bolt spacing) is probably based on the moment of inertia of the "beam," with resulting "cube function" for thickness.
3. We would expect the bolt spacing to be a function of a "power to the ¼" as the general equation for a beam with a uniform load along its length to have deflection as a function of its length to the fourth power (see Ref. [2]). That same general equation for a beam with a uniform load would also have deflection as a function of its material modulus of elasticity (E) to the ¼ power (3 ¼ = 1.3, which = 78.5/60).

Thus, as a designer, we would start with an estimation of bolt spacing at 4.0 inch (for an aluminum housing design). Obviously, we could (and should) prototype this spacing in our design and test under as real conditions as possible. Of note is that we have also made assumptions of cross-sectional area of our gasket seal areas and gasket changes in thickness as the gasket goes from:

A. Uncompressed state (before fasteners are tightened).
B. Compressed state (after fasteners are tightened down to set "stops" in the design, that is, design features near the fasteners that specifically limit the gasket from

being over-compressed. All gaskets require these "stops" to allow the fasteners to have a specified compression limit.

We could also look at similar designs where the level of ingress protection (air or water) matches what we design. If we see that 4.0 inches works for these designs, that would give us some confidence that we have certainly a chance at success.

It should also be noted that one of the factors of the *overall* design would be to have a *minimum* amount of fasteners. Thus, a 5.0 inch distance between fasteners would be better than a 4.0 inch distance (with resulting savings of fasteners and the labor to tighten those fasteners). However, the 4.0 inch spacing will increase the likelihood of the gasket design sealing under additional loads that were not a part of the calculation (like shock or thermal) and thus provide the design some margin of safety.

3.4.2   *Deflection Formulae and Maximum Deflection (from* **Injection Molding** *Magazine)*

Another problem that illustrates the relationship between stress, deflection, moment of inertia, and area is shown by a method to allow the designer to determine the stress and deflection ratios of a ribbed plate compared to an unribbed plate of the same base thickness, W (Fig. 3.4). This problem will point out the importance (and ease!) of adding a rib to the design. This rib will greatly increase the strength of a section. Ribs, such as this, are easy to add in the injection molding process, casting process, or even in standard sheet metal design. Fig. 3.4 shows two charts. For the chart marked "stress ratio (ribbed/unribbed)", the ordinate is the stress ratio and the abcissa is the rib height/base thickness. This chart shows how the maximum flexural stress changes as ribs are added to a flat plate. Each curve represents a particular rib spacing ratio, with the curve labeled ".01" representing very widely spaced ribs while other curves have more densely spaced ribs. The chart marked 'deflection ratio' is similar and shows a curve labeled ".01" representing very widely spaced ribs. More detail is available in Ref. [4].

Procedure

1. Calculate the equivalent base width, $Beq = B/N$, where

$$Beq = \text{Equivalent base width}$$

$$B = \text{Total width of plate}$$

$$N = \text{Total number of ribs}$$

2. Calculate the rib tip thickness, $t = T - 2H\,(\tan a)$, where

$$t = \text{thickness of rib at the tip}$$

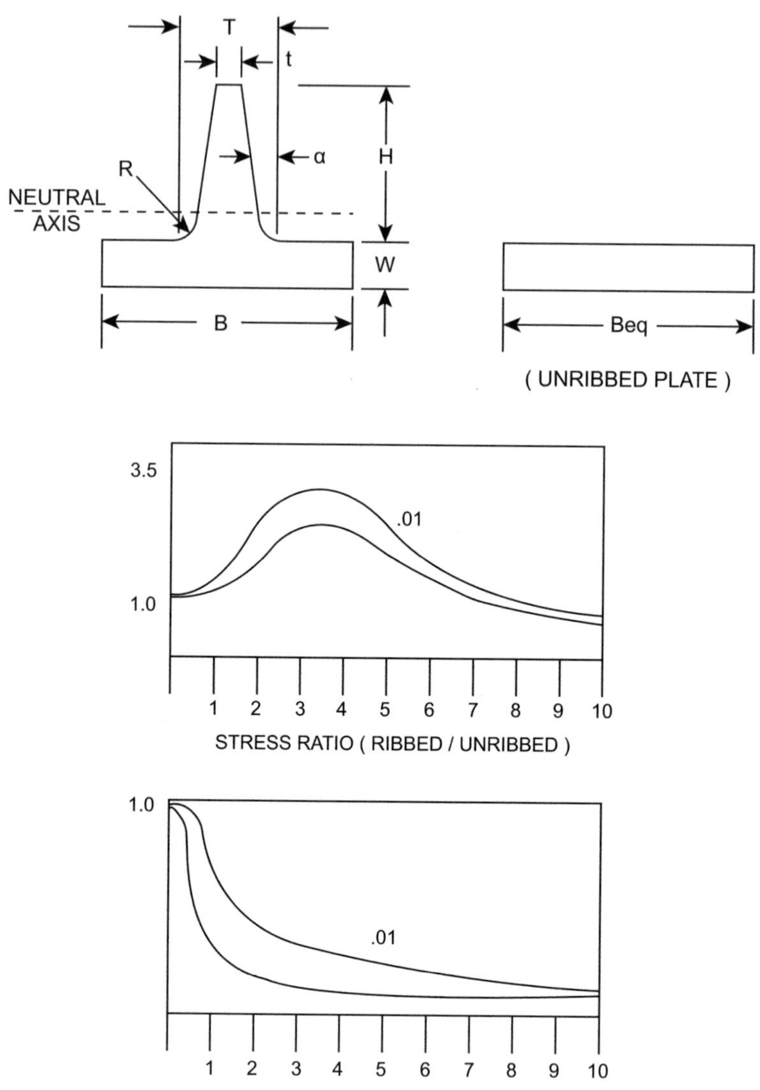

Fig. 3.4 Ribbed vs. unribbed plate

$$T = \text{thickness of rib at the base}$$

$$H = \text{height of rib}$$

$$a = \text{draft angle per side of the rib}$$

3. Calculate the cross-sectional area of the equivalent base section, $Ar = BeqW + H\,((T + t)/2)$ where:

$$Ar = \text{Cross} - \text{sectional area of equivalent base section}$$

$$W = \text{Thickness of base}$$

4. Calculate the distance from the extreme fiber to the neutral axis, $Y = H + W - (3Beq\,W^2 + 3Ht\,(H + 2W) + H(T - t)(H + 3W))/6Ar$
5. Calculate the moment of inertia of the equivalent base section, $Ir = (4BeqW^3 + H^3(3t + T)/12 - Ar(H - Y)^2$
6. Calculate the moment of inertia of the equivalent base section without ribs, $Io = (BeqW^3)/12$
7. Calculate the ratio of the stress of the ribbed plate to the stress of the unribbed plate, $Sratio = 2(Io/Ir)(Y/W)$
8. Calculate the ratio of the deflection of the ribbed plate to the deflection of the unribbed plate, $Yratio = Io/Ir$

So, we know that adding ribs to "unribbed" structure will increase that structure's ability to handle more loading. Generally, strength can be increased by adding thickness to the "general" wall thickness as:

$$\sigma = Mc\,/\,I$$

where:

σ = the stress that the member under consideration
M = the maximum moment in that member (usually a function of force times "distance," that is the "distance" from the force to the section of the member)
I/c = Z, which is a property of the section under consideration, also known as the section modulus
c = distance from neutral axis of member to "outside fiber"
I = moment of inertia (about the centroid) of the member

Thus, to increase the amount of load-bearing ability of a member, you could:

Increase I and/or decrease c (increasing Z).

The I for a rectangle (a rectangle being a common choice for a fabricated member),

$$Irec = bh^3\,/\,12,\,\text{and}\,c = h\,/\,2$$

Thus,

$$Irec\,/\,c = bh^2\,/\,6$$

where b = the length of the rectangle's base and h is the *thickness* of the rectangle.

Note that increasing the thickness (h) has a large affect due to the "squaring function."

Thus, doubling the thickness essentially makes the beam stronger by *four* times.

The above being said, doubling the thickness will increase the weight of a member (of a "standard" cross section) by *two*. This can be a "disaster" for weight-sensitive designs (which are most prevalent in the electronic enclosure industry).

However, by adding ribs, which are "intermittent" additions of thickness, the strength goes up considerably (while the weight goes up by only a small amount). The designer may be surprised to see that adding ribs may actually increase the maximum stress. Why is this? Although a rib increases the overall moment of inertia of the plate, the distance from the neutral axis to the extreme fiber of the cross section (c) can increase more rapidly for short ribs. This effect is most pronounced for widely spaced ribs.

Let's go back to the seven steps in calculating Sratio (ratio of maximum allowable stress for both an unribbed and single-ribbed design) for a very simple rib addition where the "rib" is *not tapered*, that is, $T = t$:

Width of the plate (B) = 1 inch
Single rib, height of rib (H) = 0.375
Thickness of base (W) = 0.125 inch
Wr = 0.0.125/1.00 = 0.125 rib height/base thickness = 0.375/0.125 = 3.0

$$Beq = B = 1.00\,\text{inch}$$

$$t = T = 0.125\,\text{inch}$$

$$Ar = BeqW + H\big((T+t)/2\big) = \big(1 \times 0.125\big) + \big(0.375 \times 0.125\big) = 0.172\,\text{in}^2$$

$$Y = \big(0.375 + 0.125\big) - \big((0.047) + (0.053) + (0.035)\big)/1.032 = 0.5 - 0.131 = 0.369\,\text{in.}$$

$$Ir = \big(0.0078 + 0.026\big)/12 - 0.172\big(0.375 - 0.369\big)^2) = 0.00282 - 0.00001 = 0.0028\,\text{in}^4$$

$$Io = 0.00016\,\text{in}^4$$

$$Sratio = 2\big(0.00016/0.0028\big)\big(0.365/0.125\big) = 2\big(0.057\big)\big(2.92\big) = 0.33$$

Thus, the addition of the rib to the design makes the section approximately three times stronger.

The *Injection Molding Magazine* (Ref. 7) article also compares the deflection ratio for a ribbed/unribbed section.

3.4.3   *Another Deflection Problem, This Time Snap-Fitting Hook (from* **Mobay Design Manual, Snap-Fit Joints in Plastics***)*

This problem is a great example of what an enclosure designer faces when designing a commonly used feature, the "snap-fit." Snap joints are a very simple, economical, and rapid way of joining two different components. As this eliminates fasteners from joining the two components, it is used quite frequently. The design utilizes a protruding feature from one of the parts (the "hook"), while the other part contains hole (or "undercut"). The idea here is that the hook is deflected briefly during the joining operation and catches in the undercut to complete the mating operation.

This introductory problem has been chosen as it:

A. Shows a common fastening methodology (for plastics).
B. Shows the use of common strength of materials formulae being used in a manner that utilizes the elastic nature of material, deflection being used as an advantage in a design, and optimization of cross-sectional area and uniform strain.
C. Introduces some of the aspects of designing with plastic materials.
D. Utilizes available literature from suppliers (in this case, Mobay Plastics). Instead of solving some of the more complex (yet, common) problems by first principles, the use of tabulated options and nomograms can greatly reduce the design time needed.

This calculation example is for a snap-fitting hook of rectangular cross section and with a constant decrease in thickness from h at the root to $h/2$ at the end of the hook (see Fig. 3.5). This is therefore design type 2 (see Table in Reference). General design goal is to permit maximum deformation with minimum material.

Given:

Material = Polycarbonate

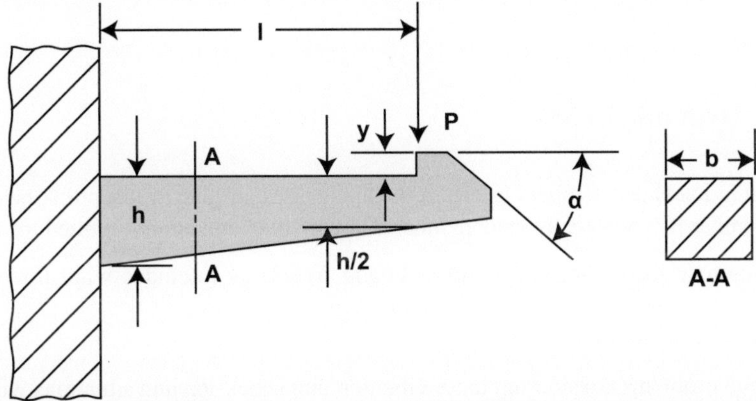

Fig. 3.5 Snap – fitting hook

Length (l) = 0.75 inch
Width (b) = 0.37 inch
Undercut (y) = 0.094 inch
Angle of inclination (α) = 30°

Find:
Thickness (h) at which full deflection (y) will cause a strain of ½ the permissible strain.

From table (for polycarbonate), ε (permissible) = 4%; therefore, ε (allowable) =2%.

From table (type 2 designs)

$$y = 1.09\varepsilon l^2 / h = 1.09 \times 0.02 \times 0.75^2 / 0.094 = 0.13\,\text{inch}$$

Deflection force (P)
From table (force equation)

$$P = \left(bh^2 / 6\right)\left(E\varepsilon / l\right)$$

From graph for polycarbonate $(\varepsilon = 2\%)$, E = 264,000 psi

$$P = \left(0.37 \times 0.13^\dagger\right) / 6 \times \left(264{,}000 \times 0.02 / 0.37\right) = 2.9lb.$$

Mating force (W)

$$W = P\left(\mu + \tan\alpha\right) / \left(1 - \mu\tan\alpha\right)$$

Friction coefficient from Table (PC against PC) μ = 0.50 × 1.2 = 0.6.
From figure, $(\mu + \tan\alpha)/(1 - \mu\tan\alpha)$ = 1.75 for μ = 0.6 and α = 30°

$$W = 2.9 \times 1.75 = 4.3\,\text{lbs.}$$

3.5   Dynamic Loads

Dynamic loads on members in electronic enclosures are due to loads that bear on the member in a nonsteady-state manner. They include, but are not limited to:

A. Vibratory loads that have amplitude and frequency (includes wind forces or inertia forces associated with earthquake ground motion)
B. Discrete shock loads

Some problems considering these vibration and shock-loading situations will be explored in Chap. 7 on "Product Environments."

Chapter Summary

In this chapter, I've introduced the EPE Designer to some of the basic considerations of the structural considerations of the enclosure. We can start this design by proposing materials for these outer hulls. Also our design disposal will be the choice of the cross section of the hull. The best choice for these cross sections and materials are made by using strength of material equations which are readily available. However, there are choices to be made among various solutions, and it will take more considerations than just structure alone to determine the best design.

Also, we have introduced a generic process for designing the structure for the electronic enclosure. This starts with looking at previous designs, determining the forces on the structure, and continues on to determining the factor of safety in our design.

From there, we looked at some examples that illustrate common problems in designing the structure. We finished with a short section on additional complications and considerations to be noted which serves as an introduction to Chap. 4.

References

Again, this chapter has been a review of structural considerations as they relate to structures encountered in the design of electronic enclosures. The reader may have many other sources of information, but the main ones I have used (over the years) are:

1. Roark RJ, Young WC (1975) Formulas for stress and strain. McGraw-Hill Book Co., New York
2. Byars EF, Snyder RD (1969) Engineering mechanics of deformable bodies. International Textbook Co., Scranton
3. Design guide to the selection and application of EMI shielding materials. TECKNIT, EMI Shielding Products (1991)
4. Injection Molding magazine, May 1998 issue, R. Cramer of Dow Materials Engineering Center

Chapter 4
Materials and Processes

Now that we have the structural foundation for the design, we'll actually start this chapter with a "return to basics." We've already touched upon the need to define and then conform to the product specification, but now we'll return to cost considerations of the design. With that reestablishment of this design "touchstone," we'll continue on with more "building blocks" that will be available to the designer to determine the best materials and processes for their enclosure parts. The choice of material and process for the individual parts that make up the assembly will get the designer also thinking about the assembly and servicing of the product (which is taken up Chap. 6).

4.1 Cost Versus Time Versus Specification

This chapter will start with a return to the basic consideration of the design, and that is an emphasis on *cost* being the deciding factor (ultimately) in the decision to make one choice over another in the design process.

A designer of electronic enclosures faces a certain "practicality" in their design in that the design must be produced on a scale that assures financial success for the owners of the company. There would be certain designs that would be considered "one-offs," where cost considerations are less important, but I'd like to address those designs that will produce assemblies (parts) that are at the very minimum, in the hundreds. I have worked at an experimental laboratory where only *one* assembly was to be produced, but again, I will not be addressing that case. Cost can even be extremely important in the case of a "one-off," such as a space satellite, but the cost of failure could then dominate the design rationale. This is also true in matters of safety or public health.

Let's further explore the above *cost* emphasis on design. There are cases where prototypes required for the final design need to be developed. These "prototypes" are

© Springer International Publishing AG, part of Springer Nature 2019

T. Serksnis, *Designing Electronic Product Enclosures*,

https://doi.org/10.1007/978-3-319-69395-8_4

certainly less cost-sensitive, as it is *time* that is usually the critical factor here. However, even though the prototype itself may not have a cost-sensitivity, the *overall* project cost is impacted in the sense that cost is sacrificed for speed *just* in the prototype portion of the project, as that time saved (by the "high-cost" prototype) results in the product being tested and approved for production in a shorter length of time, which usually translates into an *overall* lower cost (for the project).

Let me give an example where a certain aspect of the design is originally thought to be the *most* important, but it is really *cost* that turns out to be the #1 consideration. If a corporation chooses "aesthetics," that is, how a product looks and feels to the customer, as the #1 consideration, here is how that "plays out" in the marketplace. What is being decided by choosing "aesthetics" is actually a choice saying that these products will *sell more* with that look. So, the product development team's investment in "aesthetics" will actually result in increased profit to the corporation (over a product with "lesser" emphasis on aesthetics).

So, when I say *cost* in the above paragraphs, that can be also thought of as *profitability* or (increased margin), that is, lower cost = higher profits.

Time plays into this "cost picture" very much. The "time-to-market" can be a *huge* driver in product development. That is, if a certain product isn't released in some specific time frame (such as the spring planting season or the electronic show before the holidays), that can mean a huge difference to the total sales of the product. So, *coupled* with cost is the aspect of time. This leads to certain scenarios that are likely to play out in the life cycle of the product as:

1. Emphasis on *time* in material/process/manufacturability choice for the early stages of the development process
2. A "high-production" and cost-reduced product release can come occur in the later stages of the development process

All of this really still goes back to *cost* because it could have been determined (by the project management) that the overall cost is minimized by a "two-stage" product release (above). The *overall* sales, from the beginning of product release, to the end of product life, will be increased by this methodology. The concept of tooling needed for the project, and at what stage it is required, was explored in the section on Engineering Economy in Chap. 1.

Cost can also be broken down into several time frames, such as:

1. Development cost (until first shipment to customer)
2. Ongoing production cost of the product: materials/assembly/overhead
3. Service and warranty costs after production
4. End-of-life costs such as recycling

All of the costs *added together* make up the *total* cost – so minimizing cost in only one product phase doesn't minimize the *total* cost.

"Cost" is not only related to the cost of the individual part or assembly but also refers to the development (design) cost.

Another example where cost is still the #1 driver of the project is a project where *weight* needs to be minimized for the product to succeed. This is rationalized by the following logic (for this made-up scenario):

1. The specification of the product clearly calls out the:

   - Time required for the project delivery (expected)
   - Cost of product
   - Weight target of product (difficult to achieve)

2. Product is designed. Iteration #1 results in weight target exceeded.
3. Design is iterated; iteration #2 results in (slightly) exceeded weight target.
4. Time allotted to deliver product has been exceeded at this point.
5. Decision is made (by project management) to either:

A. Accept iteration #2 (deviate from original specification)
B. Move on to iteration #3, with a specified length of time needed for completion
   and notation that original delivery time has been exceeded

The above problem has its "roots" in weight minimization, but the solution is actually a matter of time (with time equating to cost). It's the cost of the project budget "overrun" that needs to be balanced with the need of product shipment.

So, again, as time is related to cost, the product designer must have these two related project aspects (time and cost) at the forefront of their design "mind space." Those two, plus "conformance" (meet or exceeds) to specification, make for an integrated approach to successful design.

Designers, if given any problem/challenge, must *always* be asking:

1. What is the acceptance criterion for the design? (How do I know I've been a success?) This usually is in the form of a specification, which may be formal or informal. The design should be working to formalize the acceptance criterion so that this is completely transparent to the project team.
2. What is the budget for the design in terms of dollars?
3. What is the project schedule for individual parts, as it relates to the entire product, and what is the "critical path" of the schedule? If the time estimated to complete the task is too short (not enough time left), other solutions such as getting more resources must be suggested as soon as known. The detail of the schedule should be such that every time-intensive activity is noted, including design reviews needed to move forward and potential issue resolving time (second iterations of design) needed.

all of the above is so important to the choice of:

- Materials of the individual parts.
- Process needed to produce the above parts.
- Assembly procedure needed to assemble above parts.
- Testing procedure needed to test above parts and assemblies.
- Quality control procedures in place to assure parts and assemblies are produced and assembled to specifications.
- Service (expected or unexpected) requirements are met.

that this will bear repeating over and over in Chaps. 4 and 5, so I'll just put the code "Cost (Chpt4)" in the text (the reader can just refer back to this general discussion as a refresher if needed).

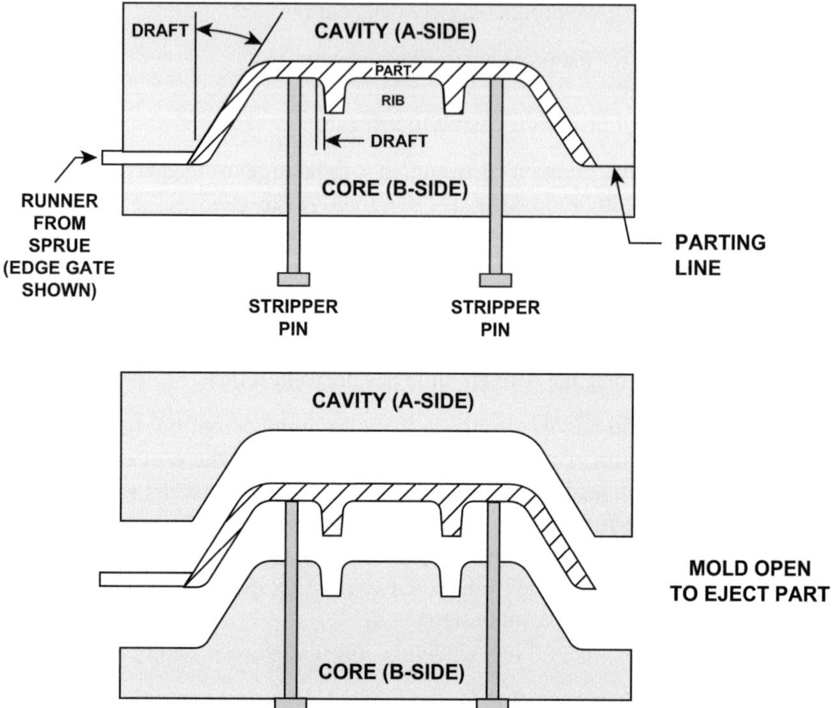

Fig. 4.1 Injection molding – mold schematic

4.2   The Designer's Mind Space

The designer has to "think ahead." When faced with designing an electronic enclosure, here are some things that go on in the mind of a designer, hopefully, all at the
same time (Cost (Chpt4)). I consider these the following questions to be "ever-
present" in the designer's head, which is why I use the term "designer's mind space"
to describe these ever-present questions:

- How big?
- How many parts are needed to accomplish purpose?
- Has this (or a slight variation) been done before? Here, or at another company?
  How has the product purpose been accomplished by the competition?
- What is the "user interface," that is, how will the customer use this product (but-
  tons/displays/lights/doors/connections for power, input, and output)?
- I am designing this portion of the product, what are the other portions that I am
  not (directly) responsible for?
- How quickly can I present some ideas that will solve the problem? How quickly
  can I prototype these ideas to check out the feasibility of an idea? Who else can
  I brainstorm with to critique these ideas?

- Once the idea has been reviewed and the prototype seems to work, what parts of the design are:
- The most risky (may not work as intended)?
- The most simple?
- The longest lead items for a preproduction run of parts, that is, what are the parts that are in the "critical path" of project completion?

For the prototype, how close is it to the production version? What testing will determine whether the prototype "succeeds" or "fails," and are several rounds of prototyping required (each round perhaps nearer to the production version)? How many prototypes does the project team need? And when?

How will I convey project progress or issues with the design to the rest of the design team? Who needs to be there for a design review?

4.3 Materials and Process Choice

Once a designer has designed a part, the designer must determine the "best" Cost (Chpt4) way for that part to be produced. The general items to be determined for *each* part are:

- Material of the part.
- Finish required for the part (see next section).
- Dimensional accuracy needed for the part.
- Process by which that part will be produced (perhaps one process for early needs, prototyping, and preproduction of parts and a different process for mature production of the parts).
- Quantity needed of the part (say, per quarter, per month, per year).
- Second operations needed for the part (beyond finishing).
- Cost requirements for the part.
- Can this part be combined with another part in the design? Essentially, what needs to be determined is whether a single (combined) part can fulfill the functionality of the separate parts (Cost (Chpt4)).
- Can the part be made symmetrical (for assembly ease)? Should the part that is *almost* symmetrical be made a more obvious nonsymmetrical part? These two questions deal with the assembly of this part and the chances of being assembled in an incorrect way. Holes or notches (superfluous) can be added to the part solely for the purpose of making that part symmetrical.

Considerations for determining material selection for a part:
The designer should choose a material that will satisfy (meet or exceed):

1. Strength requirements
2. Weight requirements
3. Reliability requirements
4. Regulatory requirements
5. Safety requirements

6. Thermal requirements
7. Shielding requirements (EMI/RFI)
8. Compatibility requirements for metals (galvanic corrosion)
9. Elastic requirements (durometer)
10. Conductive (or insulating) requirements
11. Opaqueness requirements
12. Wear requirements
13. Aesthetical requirements (touch, visual)
14. Acoustical requirements
15. Ultraviolet (UV) transmission and resistance requirements

Let's go thru a few examples and see how the 15 requirements above get determined (Cost (Chpt4)).

Example 1 Cell Phone (Outer) Case

For the choice of material, two major candidates come to mind; this would be either a metal or a plastic. Either can fulfill the 15 requirements, with these notations:

1. Metal will provide adequate EMI shielding but may be difficult to fabricate. The case needs to be smooth and elegant, which is expensive for a metal to accomplish – even a casting can require many "second operations" (such as machining or grinding) that can be labor-intensive.
2. Plastic (injection molded) will be adequate for alleviating the "smoothness" and "elegance" criteria but will need an additional EMI shielding scheme. Either the plastic case must be "metalized" or another nonaesthetic part needs to be added to the design just under the plastic case to function as an EMI shield. We also will need to investigate some safety issues with plastics (UL regulations for "flame class"). A plastic case as an exterior part could have much lower costs required for finishing (no painting required, at least that is the hope).
3. If weight (and size) is a factor (usually is with a cell phone), both metal and plastic should be examined more closely to see what may be better. Normally, ribs can be added to a plastic design to augment strength.
4. Thermal issues can be a factor in choice of material. A plastic will function as an insulator (keeping in the heat), while a metal will take the internal heat generation away to the ambient air resulting in a lower "case ambient" temperature. However, the metal case may feel too hot to the touch.

Calculating equivalent rigidity (or stiffness) of a material is shown quite succinctly in Ref. [1], where (temperature-dependent qualities):

T1 = cube root of ((E2/E1) × T2^3) = Thickness of Material 1
E1 = Flexural modulus of Material 1
E2 = Flexural modulus of Material 2
T2 = Thickness of Material 2

Let's compute the equivalent thickness of a plastic (say, SABIC Cycolac) to a 0.03 inch thick aluminum part:

$$\text{Tplastic} = \text{cube root of} \left((100/4) \times (0.03)^3 \right) = 0.088\,\text{inch}$$

Let's compute the equivalent thickness of titanium to a 0.03 inch thick aluminum part:

$$\text{Titanium} = \text{cube root of}\left(\left(100/160\right)\times\left(0.03\right)^3\right) = 0.026\,\text{inch}$$

Now, if we look at strength (all of the above are "equivalent" strengths) to weight ratios for (looking at thickness to density ratio):

$$\text{Plastic} = 0.088/0.04 = 2.2\,\text{in}^2\,/\,\text{lb}$$

$$\text{Aluminum} = 0.030/0.10 = 0.3\,\text{in}^2\,/\,\text{lb}$$

$$\text{Titanium} = 0.026/0.16 = 0.16\,\text{in}^2\,/\,\text{lb}$$

Therefore, if weight (for equivalent strength) is an issue, titanium would be a much better choice (than aluminum or plastic). This is why titanium is extensively used in aircraft (Cost (Chpt4)); however, titanium has some inherent fabrication difficulties which make it very expensive as a choice for an electronic enclosure.

Discussion on molding plastics is in a following subchapter.

No decision will be made here of using "metal vs. plastic" for the exterior case of the enclosure. Enough ambiguities have been raised as to warrant a much more in-depth analysis, and we need to move on to another example of material choice. The Apple 5s phone case was made of (machined) aluminum, while the later released Apple 5c phone case was made of plastic. The plastic was made EMI compliant with the addition of a metalized coating in some areas and metal shields in other areas. The plastic version was thought to "bend" in the shirt pocket much more easily, but not enough to stub sales. Machined aluminum was a much "sooner to market" choice over molded plastic, as long lead time tooling was not required. Also, changes can be made much more adroitly with a machined piece than a molded part due to the longer time needed to modify the tool and test the new change. Again, each design will likely be unique to material and process choice – this is an example of something the mechanical engineer (working with other teams in the corporation) can make to the overall product design.

The optimum choice between plastics and die castings requires careful analysis of all product requirements.

Aluminum, magnesium, zinc, and zinc-aluminum (ZA) die castings are often preferred over plastics in electronic enclosures where strength, stiffness, and minimum packaging space are required. They often eliminate the need for inserts to received threaded fasteners. Their thermal conductivity often eliminates the need for cooling fans, which is essential in battery-operated portable electronic equipment. Die castings are also preferred over plastics where strength and rigidity are required and for medium to large size decorative components operating at elevated temperatures.

EMI and RFI (electromagnetic interference and radio-frequency interference) shieldings are inherent with die castings as they are metal. Plastic parts need a layer of metal shielding to provide EMI/RFI compliance. These shielding additions for plastic parts (painting, coating, resin fillers, metallic barriers, multilayered electroless nickel plating) can have performance problems and quality control issues.

Aluminum die castings are frequently chosen over plastics in designs that are subject to continuous pressure, particularly at elevated temperatures, and hand tools such as drills that require minimum weight, rigidity, and good surface quality. Magnesium die castings are frequently chosen in applications that require minimum weight combined with strength, stiffness, and minimum packaging space. Components for high-speed printers, which require rigidity at minimum weight; cases requiring mounting features, high-quality surface, and impact resistance; and decorative trim pieces are frequently produced by magnesium die casting.

Plastic components that are subjected to continuous loads, such as a part being used to environmentally seal, often require support from metal stampings to develop stiffness and creep resistance.

Also, it should be noted that even a "simple" material such as aluminum is much more complicated to actually specify in an engineering drawing. Many grades and alloys of aluminums exist, either "commonly" or "exotically," and this text is not going to address that complication, only to say that much attention is deserved to actually specify the specific "grade" of the material. Picking one grade over another is a function of that particular grade's ability to (Cost (Chpt4)):

1. Possess the 14 characteristics (stated above) of the design.
2. Be available to any of the supply chain members for meeting delivery times.
3. Be fabricated in a repeatable manner with specified quality control measures.

For example, if I wanted a part to be made of stainless steel, items to be considered would be:

1. Choice between chromium nickel stainless steels (austenitic) and straight chromium stainless steels (martensitic). Chromium nickel stainless steels are Types 20X and Types 3XX. For example, Type 302 is (from Ref. [2]):

   - The basic 18% Cr. 8% Ni, analysis, possessing excellent corrosion resistance to many organic and inorganic acids, and their salts at ordinary temperature.
   - Has good resistance to oxidation at elevated temperatures.
   - Can be readily fabricated by all methods usually employed with carbon steels.
   - Cr-Ni grades are nonmagnetic in the fully annealed condition and cannot be hardened by conventional heat treatment.

   In Ref. [2], Tables of Information (for the various grades of stainless steel): chemical composition, physical data, and the mechanical properties (both in the annealed and heat-treated states) are shown.

2. *Availability* must be considered for the candidates (choices) of material. In this example for stainless steel, by checking a reference such as Ref. [3], Type 302 *is not readily* available. Types 304, 309, 316, and others are available in stock (from Ryerson, a large resource of stainless steel stock). For example, Type 316 (cold rolled, annealed, pickled, 2B and 3 finish. Spec: QQ-S-766), ASTM-A240, is available in 16 gauge (0.060 thick) × 30 inch × 96 inch size stainless steel sheet.

3. Material choice should be made with fabrication technique in mind. If the part is envisioned to be turned on a spindle machine (lathe or mill), then a free-machining grade needs to be considered. If the part is to be welded, various grades may or may not be very weldable.
4. As usual, information about the material choice can be utilized using experience from:

   - Your codesigners within your own group (or other corporate resources),
   - Your supply chain fabricators

Once the material choice is made, it must be fully specified, that is, specified so that it will be unambiguous on the part specification. Materials and finishes are usually specified to some standard such as ASTM, MIL-standard (US government), or international standards such as ISO/IEC. There must also be in place some methodology of assuring (thru standard quality control procedures) that the material specified is the material being fabricated into the final part.

Example 2 Cover for a LED
Let's take another example of a material choice and the "thought process" that one would go thru to make a rational choice among candidate materials.

How about the need for a clear (optically transparent) part that will cover an LED? This part is envisioned to be flat and not needing much structural strength, basically supporting its own weight and preventing the user from "poking" the LED with their finger or a pencil. Let's name the part "LED *window*" for now. The piece will "seal off" the LED (from moisture) and is not generally replaceable (must last the product lifetime without being serviced).

Immediate questions that should come to the designer's mind are (Cost (Chpt4)):

1. How many are to be produced, what is the schedule for part production (prototype/preproduction/production), and at what general cost? The cost is usually not specified up front in the design, that is, it may be something ambiguous such as "as cheap as possible." As the design proceeds, certain choices in the design may increase the cost of the part, so trade-offs with alternatives are usually helpful.
2. Will it be designed to fit into a recess or "stand out" from the enclosure? How far away will the window be from the LED? These are the general "geometry" issues of the design. What are the aestetic considerations of the window?
3. The light from the LED shines thru this part. Is there a need for the light (from the LED) to be "diffused" – what kind of look do we want aesthetically? How does this part look in the "overall" design? Is it in the user's view all the time or only occasionally? Is "clear" the right choice of color, or is there another color that is wanted (red/green/amber/blue)? What color is the light from the LED?
4. Go thru the 15 requirements for the material from the above characteristics. For example, how is the part (right now, a nonmetal) going to adversely affect the EMI shielding requirements? This could possibly lead to the new part being as small as possible to create a minimum hole for EMI.

5. How will the LED *window* be assembled to the enclosure? Can it be assembled without any extra hardware? Candidates for assembly may be the use of ultrasonics, tape, or adhesive to bond the piece into place.
6. Are there multiple (more than one) "windows" in the overall design? Look for commonality possibilities with these windows.
7. Material candidates (with some issues for that choice):

   - Lexan plastic: Molded or cut from sheet? If molded, can we see mold flow lines? If cut from sheet, it probably will be flat (no other geometric features), while molding allows a design feature to be added (such as an ultrasonic weld bead). How thick? Is there an "anti-scratch" or "antiglare" requirement (and how would that be solved?)? If molded, can the *window* be assembled to the enclosure at the enclosure fabricator? Plastic will have some safety considerations such as being a burn hazard or having sharp edges.
   - Glass: Hardest and toughest material candidate, probably the most expensive (Cost (Chpt4)).
   - Polyester film as part of a label: Usually under 0.010 inch thick. Label could incorporate other information on the enclosure including identifying what the LED functions are.

8. Present to the project team with whatever detail is warranted. This could be price/time estimates, prototypes, or sketches to base going forward with a design choice for material.

## 4.4	Finishes and Coatings

All of the choices made when selecting a material (previous selection) are directly "coupled" with the choice of finish for that material. Practically all engineered parts *need* a finish. There would be some exceptions to this, for example, a "sculpture" (artwork) or building façade that is intended to corrode (and have a "corroded" look). The designer will be specifying *both* a material *and* a finish to every part they design.

Finishes (including coatings in the broad sense) are required to:

1. Retard corrosion in storage (from fabricator, to assembler, to customer) or in final usage by the customer.
2. Provide anodic protection when metals are in contact with dissimilar metals. Basic to this, dissimilar metals in contact must have adequate protection against galvanic corrosion. This is accomplished by interposing inert material or that which is compatible to each. The table below lists similar metals by groups. Contact between a material of one group and another material of the same group shall be considered as similar. Conversely, a critical electrolytic stage is set (requiring only humidity as the agent) whenever materials of different groups are in intimate contact. (This is particularly disadvantageous when either magnesium or aluminum, unprotected, is in contact with any other metals of different groups (Table 4.1).)

Table 4.1 Material groups

Group I	Group II	Group III	Group IV
Magnesium alloys	Aluminum	Zinc	Copper
	Aluminum alloys	Cadmium	Copper alloys
	Zinc	Steel	Nickel
	Cadmium	Lead	Nickel alloys
	Tin	Tin	Chromium
	Stainless steel	Stainless steel	Stainless steel
			Gold
			Silver

Note that the uses of lead, cadmium, and, in fact, all finishing materials have very serious environmental concerns. Many of these materials are banned or limited by various legislative statutes and laws. Please see the RoHS requirements as an example of these international laws that limit the use of one or more of the above finishes.

3. Appearance (aesthetics).
4. In the case of bonded connections, the protective coating will actually be omitted (masked). For such areas, moisture entrance must be prevented by forced ventilation or adequate sealing.

Finishes are usually listed into three main types: (See Ref. [4, 5])

- Chemical: those finishes resulting from chemical reactions on the surface of the metal.
- Electroplated: those finishes consisting of a film or plate deposited on the base metal of electrolytic action.
- Organic: finishes consisting of an organic coating over a base material, applied usually by brushing, dipping, or spraying.

Coatings can be applied by:

- Sprayed metal: A thin layer of metal is sprayed onto the surface for several purposes. Examples are aluminum being used for corrosion and heat resistance or copper for electrical conductivity.
- Powder coating: A dry painting process in which powder particles are applied directly to the surface to be coated without the use of solvents or water. Either thermosetting or thermoplastic powders are used. Parts are electrostatically powder sprayed at room temperature and then heated above the melting point of the powder to attain a fused surface finish.
- Electrodepositing: A thin coating can be deposited electrically to improve appearance, increase electrical qualities, and increase resistance to wear, corrosion, or specific environments.
- Ceramic, cermet, and refractory: Fixed porcelain enamel frits and refractory materials are used as corrosion-resistant coatings and also for color appeal and decorative effect.

- Hot dipping: These coatings, used principally on steel, cast iron, and copper, provide corrosion resistance at low cost. Materials used are aluminum, zinc, lead, tin, and lead-tin.
- Immersion: These coatings can be applied to most ferrous and nonferrous metals, with a few exceptions. Materials used are nickel, tin, copper, gold, silver, and platinum. Examples of use are for conductivity, facilitation of soldering, and brazing.
- Diffusion: These coatings are produced by the application of heat while the base material is in contact with a powder or solution. Most diffusion coatings are intended to obtain hard and wear-resistant surfaces and to increase resistance to corrosion.
- Vapor deposited: This is depositing vaporized metal in a vacuum chamber, where it then condenses on all cool surfaces. Most metals and nonmetals can be used as base materials to be coated. Examples are mirrors and optical reflectors, metalized plastics, lens coatings, and instrument parts.
- Organic: These consist of alkyds, celluloses, epoxies, phenolics, silicones, vinyls, rubbers, and others.
- Chemical conversion: These are chemical coatings which react with the base metal to produce a surface structure that will improve paint bonding, corrosion resistance, decorative properties, and wear resistance. Phosphate, chromate, anodic, and oxide coatings are common.
- Rust prevention: These are oils, petroleum derivatives, and waxes that form a film which will resist attack, principally from industrial and marine atmospheres.

Finishes or coatings that will be applied to engineering materials are usually called out on the (part) documentation with a MIL *specification* (MIL SPEC) Cross Reference. This is mainly done because most (common) finishes already are standardized and calling out an existing specification:

1. Saves time in that standards already exist that are "universally" accepted.
2. Suppliers already have these processes in place to economically produce these finishes.
3. The finish can be checked (verified by the specifier) using acceptable, in-place quality control procedures.

For example, a chemical film for aluminum may be called out to comply with a chemical film per MIL-C-5541.

An example of a callout for engineering documentation for a material and finish is:

- Material: 16 Ga. (0.060) 1010 CRS (with CRS standing for cold rolled steel and the "1010" a shorthand for AISI M1010 Steel). AISI M1010 Steel is a low carbon, general purpose merchant quality steel, featuring economy plus formability and weldability.
- Finish: Zinc plate clear chromate per QQ-Z-325. Class 2, Type II (with QQ-Z-325 being a common MIL SPEC for zinc and the "class" and "type" specific choices for attributes such as minimum thickness).

One of the most extensive finishes for metals is paint. Painting can be very much a complicated process. Difficulties occur with:

- Surface preparation
- Color matching
- Identification and control of defects

Ref. [9] is included as further information.

The material and finish callout on the engineering documentation should be unambiguously stated. It is a good idea to see how your company usually calls out these common materials and finishes, check with the supplier as to how the callout fits in with their processes, and what the quality control procedures for "guaranteeing" the material and finish will be handled by both at the supplier and at incoming (to your facility) inspections. Sometimes this is handled with a material/finish "certification" that is supplied to the incoming inspection by the supplier.

4.5 Punching and Forming Metals

The basic processes by which metals are punched, notched, formed, and bent has changed in recent years. Old fabrication techniques like (non-CNC) lathes, milling machines, and drill presses have been relegated to "garage shops" or "quick-turn" prototype shops (of the past). Contemporary fabrication is done on CNC (computerized numerical controlled) multi-axis machines or high-speed strippet punch presses. Your CAD file is electronically transferred to the shop, the shop "converts" the file as input to their fabrication machine, and the machine creates the part (in specified quantity).

A strippet punch press takes a flat piece of stock and positions a rotating turret (pre-loaded) with round punches, rectangular punches (for cutting the periphery), and just about any shape punch (customized or already in the shop catalog), over that piece of stock. The table that the stock resides on moves in x and y, and the turret rotates to put the proper punch in place. A flat piece of metal can be completely punched out with a complex design in minutes. Specialized "punches" that punch and form louvers can also be programmed on the strippet. Tolerancing (which we will expand upon in the following section) can be very tightly controlled with CNC as there is no "manual" setting of dials, stops, or machine feeds/speeds.

Multi-axis spindle machines are commonplace in today's fabrication environment. The term "5-axis" is typically referring to the ability of a CNC machine to move a part or a tool on five different axes at the same time. 3-axis machining centers move a part in 2 directions (X and Y), and the tool moves up and down (Z). 5-axis machining centers can rotate on two additional rotary axes (A and B) which help the cutting tool approach the part from all directions.

As with all fabrication techniques, the more the designer is familiar with the machine and machining process, the better the design will be Cost (Chpt4).

I've included a section on Sheet Metal Practices in Appendix. These show very common practices of bending and punching metal that sheet metal fabricators use and are commonly found in corporate drafting standard manuals.

4.6 Molding Plastics

A designer of electronic enclosures must have solid knowledge of the plastic molding process and how to design plastic parts. Most engineering degrees do not place much emphasis on this skill, so it is likely attained and honed while on the job. There is much in the literature on the proper design of plastic parts, and I'll provide some good references on the subject, a lot of which come from the plastic suppliers (of raw pellets) themselves. The plastic suppliers themselves have learned a great deal over the years on the use of their resins, and it benefits everyone to share that experience. Also, Ref. [6], by Glenn L. Beall, can be considered a "bible" as Mr. Beall is a well-known expert in this field. I will be providing a list of the "top ten" guidelines for plastics design, but these will be just the highlights of an extensive list that each designer can add to.

The probable #1 capability of the plastic part designer is understanding the tooling that will used for their parts and having an understanding of what options are available with injection molding tooling. With an understanding of injection molding tooling, the following six concepts will help in the design of injection molded parts (see Ref. [7] for additional information):
(See Fig. 4.6)

1. The idea of draft needed to eject the part from the mold.
2. The location of the main gate (or "sprue") that will "inject" plastic into the mold. This gate location (and subsequent need of "degating") will be a large consideration for the cosmetics required for the part. Gating can be done at the edge or either the core (reverse gating) or cavity side of the mold.
3. The idea of "mold flow" needs to be well understood. The need for *radiused corners*, generally an *equal thickness* design, and *limitations on rib heights* are highlighted. As the melt cools, it turns from a viscous liquid into a semisolid and, eventually, into a solid part. It is more difficult to fill the areas of the part the furthest away from the part gate.
4. Stripper bar locations are shown, again, highlighting the cosmetic surface need.
5. If "undercuts" are required by the part, the tooling will show how this is to be achieved (and how much this can complicate the mold).
6. The mold "parting line" as shown in the tooling is a reflection of the part design and thus shows the difficulties with those designs.

The above tool design features are learned from experience. In all cases, the part design needs to be reviewed by the tool designer (usually resident or contracted by the molder), and the part designer should get assurances that their design is "moldable" in a straightforward manner. Both the part design and the tooling need serious review by the designer and the project management team, as mold tools have:

1. A large capital expense (K$)
2. Very long lead-times (they need weeks to complete)
3. Difficult (both in time and money) revision processes

One of the largest advantages of a molded plastic part is that it needs less second operations to become a finished part (as compared to a "like" metal part). Usually, the molded plastic part does not need cosmetic painting on the exterior. Most (customer visible) plastic parts end up with a molded-in texture on the exterior which is usually achieved by etching the mold with a "texture pattern." Note that this mold texture results in undercuts to the part (which is achieved by adding a small amount of draft to the part).

Some common second operations are needed to a molded part (these are operations done outside of the molding operation and usually done by the molder (or contracted out by the molder)):

1. Holes or cuts needed to the parts that were not deemed doable in the mold itself – sometimes holes are easier to be added by a second operation rather than tooled as part of the mold. Thus, drilling or tapping operations may be added.
2. Machining may be required to repair the "gate mark" if gate is in a cosmetic area. Gates may also be removed (by hand) if the gate area is not a cosmetic surface (and that surface doesn't require machining).
3. Inserts (metal, threaded fasteners) may be ultrasonically placed into the part (or these inserts can also be molded in).
4. EMI shielding can be applied to the plastic part. This takes all sorts of forms:

   - Painting
   - Plating
   - Bonding of metal shielding

5. Bonding operations (ultrasonic or adhesive) can fuse one or more molded parts together.
6. All sorts of decorating (silk-screening, painting) can be done. Some in-mold operations can be done to incorporate graphics in that manner.
7. All fixturing needed for any of the above operations need to be clearly identified, costed, and placed on an accountable time schedule.

All of the above second operations need to be clearly called out on the finished part documentation which would include any quality control acceptability criteria. These second operations can add to the piece price of a part in a very *significant* way, so they need to be a part of the total design process, with alternatives to that design clearly presented by the designer. The second operations need to be discussed in great depth with the molder to see that they are achievable in the most cost-effective manner.

Choice of plastic for the part: The designer must review all of the 15 characteristics previously listed (for *any* material) to decide what resin to choose. Creep data (long-term viscoelastic behavior) is also important for plastics. Some other unique aspects include:

1. There are many molding processes: Injection molding/blow molding/compression molding/thermoforming (pressure or vacuum forming)/reaction injection molding (RIM). I've been mainly addressing the injection molding process.
2. Snap-fits require materials with specific limits on stress vs. strain.

3. Data exists for the common choices of plastic with the plastic pellet suppliers (See Ref. [8]). These suppliers can also review the design and possibly indicate if a customized solution is needed for your application. The suppliers are a large reservoir of information on materials and molding.

As previously "promised," here are my top ten recommendations for proper plastic part design. Most designs can be analyzed with a mold-filling "program" that is resident in-house or at the molder. These programs take the CAD geometric data of the design, the proposed material, and determine an optimization process to properly fill the mold:

1. All attempts must be made to maintain a *uniform wall thickness*. Thicker areas will likely result in a cosmetic defect known as a "sink." Ribs with large draft can result in a thick wall section where the base of the rib meets the main wall (this can limit rib "height"). This is also a common occurrence with screw bosses where the boss base meets the main wall. Screw bosses also incur an additional complication as the insert in the boss requires a wall thickness needed for structural requirements. Where it seems that a larger (than nominal) wall thickness is needed – look to "core out" the large area with an added feature in the tool that will result in a (more) uniform wall thickness. Always think of the part in *three* dimensions. Commonly, the design is a series of 2D sections that are used to accomplish a purpose. However, plastic parts can have contours that result in curved areas that result in areas of nonuniform wall thickness.

2. *Radius* all transition areas for several reasons. The melt flow is easier with these radiused sections. Also, sharp corners can only be produced by mold tooling which, in itself, would be sharp. These sharp edges are difficult to maintain over the life of the tool. Inside radiused areas should "match" outside radiused areas, that is, inside radius + nominal wall thickness = outside radius (of part). The ratio of inside radius to wall thickness (R/T) should be as high as possible to reduce the stress in the cross-section, R/T > 0.75 is recommended. Some notes on plastic part drawings include a statement such as "Radius All corners 0.030 Radius or as Specified"; however, care must be taken as with any note intended to cover all occurrences on the drawing.

3. *Selection of tool maker/molder* is *critical* to overall satisfaction with the part. Other factors (besides "cost") enter into this selection, such as:

   - Quality of tool – selection of material and heat treatment of the tool.
   - Delivery schedule of tool – first article and production quantity parts.
   - Communication path openness – how well supplier can communicate with the part designer, the communication of suggestions about the part design, and any ongoing issues that may arise. Regularly scheduled milestone publication is essential. A face-to-face discussion can help communication, but the Internet can also be used effectively.
   - Total cost of part – includes delivery (molded/boxed/shipped), any fixturing required for the part, secondary operations, and quality control.
   - Supplier chosen from current approved suppliers – the cost of "qualifying" a new supplier must be considered. Also, problems with current approved suppliers are already known, while unapproved suppliers likely mean unknown problems.

4. Look to *reduce the amount of secondary operations* required for the part. Look at the trade-offs involved with tooling slides. Features in the part that can be produced by adding slides to the tool (or revolving inserts in the tool) offer a way to reduce second operations. However, these slides or revolving inserts make the tool more expensive and complicate the molding operation. For example, if a large thread is needed in the part, these can be formed within the tool by including a revolving portion of the tool (that "unthreads" to remove the part from the tool), thus eliminating a (secondary) threading operation (done outside of the tool). However, it may be less expensive to tap the threads as a secondary operation than to include the threads in a more complicated tool (that has "moving" parts). Certain holes in the part can be formed in the tooling by what is known as a "cross shutoff." This cross shutoff "pierces" a side wall with pieces matched on an angle. However, the negative to the cross shutoff is a "stepped" parting line in the tool that slightly complicates the tool. Again, discussions with the tool vendor/molder are invaluable for deciding on trade-off choice.

5. *Draft* should be as generous as possible, without compromising uniform wall thickness or the design. Zero draft is possible in some areas – please talk to the molder! (For example, texturing actually produces "undercuts" that don't need side pulls if the draft allowance allows the part to "spring" away.) How much draft that is required is a function of:

   - Material
   - Shrink
   - Part design need
   - Roughness of texture

   Minimum draft on ribs allows higher ribs while maintaining relatively uniform wall thickness where the rib meets the wall.

6. Be liberal with the amount of *reinforcing ribs* in the design. This is one of the advantages of using a molded plastic part in that part-strengthening ribs can be added with very little overhead to cost. These ribs increase stiffness and control melt flow. Patterns of ribs, such as circular and rectangular, should be explored. Connected ribs, to each other, or side walls can increase overall stiffness of the part.

7. Be careful of exactly where the *"weld lines"* will occur on the part. Weld lines occur where the material flows within the part and meets (itself) on the opposite sides of where the plastic originally enters the part (at the tool gate). These weld lines result in structurally weak areas of the part (as the material has cooled at this juncture of two meld paths). Also, the weld line shows up as an actual "line" on the exterior of the part resulting in an aesthetical issue that must be addressed.

8. Think of areas in the part design which will result in areas that produce a *very* thin section of the tool itself (or even a bent core pin). These thin tool sections can be difficult to maintain, and breakage can result to delay part molding. Always look to add "sections" to the mold which can be easily removed. These sections (*tool inserts*) can be utilized for either:

   - Easy maintenance
   - Easily producing another part variance, for example, a part with a hole and a part without a hole

For example, it is very common to put the part revision level on a tool insert so that, as the revision level changes, only the (small) insert needs to be modified.

9. Consider *adding features* into the design which will help align (or structurally help) a mating part. For example, adding a round recess in a part that can mate with a part that has a round positive pin can help the overall design of the mated pairing of these two parts.

10. Consider where the *mold stripper pins* will be placed, and "how many." The stripper pins will help eject the part from the mold after the part has (just about) solidified. However, they will usually leave a mark on the (hopefully) non-cosmetic side of the part. These locations should be reviewed by the designer to make sure they are optimally placed. For example, the stripper pins should not be placed where a gasket will be in the design (as the gasket surface will not be flat).

11. (Yes, the 11th listing in a top ten.) Search both the literature and look at other plastic parts to gain knowledge. Over the years, much "tribal knowledge" has been acquired in the areas of:

- Snap-fit design
- Fastening processes (assembly methods) and fasteners used for plastic parts
- New material blends
- In-mold decorating

The *general* plastic design process follows these six steps: Correct revision control of the documentation is assumed here – see Chap. 12 on documentation. Design review is broadly meant to include any and all people that should be on the review cycle for both the individual part and the overall project (should include in-house reviewers, contractors, and suppliers):

1. Part is designed. Design is reviewed.
2. Bids for tooling are placed. CAD files sent (as appropriate). Design is reviewed by tooling vendors and resin manufacturers. Any changes to design are again reviewed. Schedule for tool development and first article run are agreed to. Plan for first article acceptance is agreed to. All pricing for part known (including fixturing required) is agreed to.
3. Final design is approved; tooling purchase order is placed to a particular (and agreed to) revision level.
4. Tool maker inputs CAD file and adds correct shrink factor (parts will shrink out of the mold) to assure correct part geometry.
5. Meetings regularly scheduled to review project milestones. Updates are given to the project management. Any changes to part and/or time schedule are clearly broadcasted.
6. First article is produced and reviewed. Modifications considered including ramifications to project schedule. Preproduction run of parts scheduled as appropriate.

4.7 Casting Metals

Casting metals has a lot in common with the previous section on molding plastics; however, there are some clear distinctions. First of all, as in above section where there are actually quite a few processes for forming plastics, the same can be said for the casting process. There are many casting techniques. The casting techniques generally vary by:

- Metal cast
- Size range of part normally cast
- Tolerances expected to be held by process
- Cost of tooling
- Part price
- Surface finish expectation
- Minimum draft recommended
- Normal minimum section thickness
- Ordering quantity
- Normal lead time

The common casting techniques are described below (Ref. [10], and additionally in Ref. [11, 12]):

- Die casting: Molten metal is injected, under pressure, into hardened steel dies, often water cooled. Dies open and castings ejected.
- Permanent mold: Molten metal is gravity poured into cast iron molds, coated with ceramic mold wash. Cores can be metal, sand, sand shell, or other. Molds open and castings ejected. New LPPM method pressure pours with up to 15 psi.
- Investment (lost wax): Metal mold makes wax or plastic replica. These are sprued, then surrounded with investment material, baked out, and metal poured in resultant cavity. Mold is broken to remove castings.
- Plaster mold: Plaster slurry is poured onto pattern halves and allowed to set; then mold is removed from pattern, baked, and assembled, and metal is poured into resultant cavity. Molds are broken to remove castings.
- Ceramic mold: Ceramic slurry poured over cope and drag patterns, allowed to set, then molds removed from pattern, and baked (at 1800 °F), producing hard, stable molds. Molds assembled with or without cores and metal poured into resultant cavity. Molds are broken to remove castings.
- Graphite mold: Similar to ceramic mold, except graphite mold used instead. Core pins are usually steel.
- Resin shell mold: Resin-coated sand is poured onto hot metal patterns, curing into shell-like mold halves. These are removed from pattern, assembled with or without cores. Metal is poured into resultant cavities. Mold is broken to remove castings.
- Sand casting: Tempered sand is packed onto wood or metal pattern halves, removed from pattern, assembled with or without cores, and metal is poured into resultant cavities. Various core materials can be used. Molds broken to remove

Table 4.2 Table of casting techniques (from [6])

Technique	Metals	Size range	Tolerances (inch)	Surface finish (rms)
Die casting	Al/zinc/Mg	<2 ft^2	0.001–2	32–63
Permanent mold	Al/zinc/brass	Oz – 100 lbs.	0.015 basic	150–250
Investment	Most castable	Oz – 150 lbs.	0.003 basic	63–125
Plaster mold	Al/zinc/brass	<3 ft^2	0.005 basic	63–125
Ceramic mold	Most castable	<350 lbs.	0.005 basic	80–125
Graphite mold	Zinc or zinc-Al	Oz – 10 lbs.	0.005 basic	63–125
Resin shell mold	Most castable	<4 ft^2	0.008–10 basic	125–350
Sand casting	Most castable	Oz and up	0.03 basic	150–700
Metal injection mold	Ferrous	Under ¼ lbs.	0.003 basic	45
Powder metal	Ferrous/SS/Al	20 in^2	0.004 basic	16–90

castings. Specialized binders now in use can improve tolerances and surface finish.

- Metal injection molding: Very fine metal powder combined with binder material is injected into die. Part is ejected, the binder melted or dissolved, and vacuum sintered, resulting in 94–99% theoretical density.
- Powder metal: Metal powder is compressed in die barrel between moving upper and lower punches. Lower punch ejects part which is then sintered and sized if close tolerance is required.

The Table 4.2 below summarizes the above differences for the various casting techniques.

Casted parts are *generally* designed with the similar constraints of injected molded parts, that is, constant wall thickness and radii required at all corners, but there are some important differences. As most casting materials are more brittle, they are more prone to developing stress in sharp corners, so larger radii are needed. Sink that occurs with injection molding is not a huge problem with castings; however, parts still need to be cored to keep a constant wall (and save weight).

The casting process, although similar to injection molding, results in unique problems that must be addressed:

A. Porosity Castings are not homogeneous. The "skin" of castings (the outside layer, perhaps 0.020 inches thick) is relatively smooth and continuous. This results because the mold surface itself is relatively cool as compared to the "molten" flow of metal, and a skin forms on this outside edge of material that is dense. Right below this skin can be an area of air (gases) trapped within the metal that is relatively "porous." If one takes a machine cut on the outside surface, an area of porosity (uneven metal) can be exposed. Machine cuts are sometimes taken on the outside of a cast part, as the tolerance on any particular dimension can be quite high. For example, an outside dimension that is nominally designed to 4.000 inch may be cast at +/−0.010 inch. Thus, if the casting ends up at 4.010 and 4.000 inch is needed for the design, 0.010 may be needed to machine off of the casting. Machining 0.010 off of the part may take away the hard skin, revealing the underlying porosity.

Also, an "as-cast" surface may not be "smooth" enough for the design if the design needs a low surface roughness (as for a surface being used for a pressure seal). If the "as-cast" surface roughness is 500 µ-inch and a 32 µ-inch surface (for example) is needed to provide a sealing surface (with an O-ring), the surface in question would require machining.

Impregnation is sometimes used to develop pressure tightness and smooth surfaces in die castings. Systems employing anaerobics and methacrylates are currently used when impregnation is specified. These systems produce sealed castings ready for pressure testing.

Porosity, in castings, also creates a serious quality control issue. As porosity can be the result of the casting process (proper settings of temperature and pressure, plus feedstock material consistency), porosity can occur in a part at some point even though some parts in the batch were without porosity. If porosity occurs in a part, at a critical cross-section under load, the part may fail. Thus, porosity must be screened for, and the process must be controlled and monitored. If the part was a "mission-critical" part involved in safety or of a national defense need, the part would need some methodology to be employed to assure the part strength.

B. Secondary Operations Required Castings usually require some "large" operations after the casting process to make them ready for final assembly. These operations can be divided into machining and decorative/corrosion preventative coatings.

Similar to an injection molded part, the area where the material "enters" the part (at the "sprue") must be removed. That is, a machining operation is required to remove the vestige of material that remains with the part, from the injection sprue. This machining operation may be as simple as hand clippers to remove the vestige but may be more elaborate if this vestige is on a cosmetically visible surface. Unlike an injection molded part, the casting process usually also involves a "trim die" to remove material at the die parting line that either gets out as "flash" or is intentionally allowed out to help fill the entire part with solid metal. This "trim die" can consist of a press plus fixture to control the trimming process or be rather "uncontrolled" (such as a grinding wheel done "by hand"). "Flash" occurs due to tiny mismatch between the upper die and lower die. Where the upper and lower die "come together," the pressure of the casting process forces some small amount (0.001–0.005 inch) of material to be "squirted out," and as the dies get more used, this situation gets worse. A trim die (and fixture) can cost thousands of dollars to avoid trimming by hand. So, this is another cost/time decision to be made substantiating any tooling expenditure.

A variety of surface treatment systems can be applied to die castings to provide decorative effects, corrosion protection, or increased hardness and wear resistance. It is recommended to seek specific information on surface treatments from suppliers and references, as these vary by cast material and purpose of the surface treatment. For example, for cast aluminums:

- Decorative finishes can be achieved by paint, polish/epoxy, plating, and powder coating.
- Environmental corrosion barriers can be achieved by paint, anodizing, chromate, and iridite.

- Fill and seal surface/subsurface can be achieved by impregnation.
- Improved wear resistance can be achieved by hard anodization.

No designer of castings should be without the reference material of Ref. [13]. The NADCA product specification standards for die castings is a "bible" showing required background information including:

- Process and material selection
- Tooling for die casting
- Alloy data
- Engineering and design (specific to die castings)
- Quality assurance
- Commercial practices

4.8   Dimensioning/Tolerancing

4.8.1   Choice of "Nominal Dimension"

This may be an unusual section in this book, but I believe it will help greatly in the journey to designing great parts. I'd like to start with some basics about why we choose "the numbers" like we do, and how that choice of numbers leads to a design. How the designer ends up dimensioning a part can lead to years of trouble-free assembly and very happy customers. If this is not done correctly, the parts are in a rather constant state of "not fitting" and "out of tolerance," and assembly line stoppage will occur. In some sense, this section of the book is a precursor to Chap. 6 on assembly and serviceability, as properly dimensioned and toleranced parts will lead to very smooth manufacturing assembly of those parts.

Some comments about the English system of units vs. metric system of units are also appropriate (see separate discussion). I even want to start this discussion with how I actually started design "parts" (which were actually tooling fixtures and jigs to perform machining or welding of parts). This was before CNC'd parts, computers, and CAD were a part of the "standard toolbox." We actually designed parts with *fractions* in mind. That is, we tried to design parts using the common markings of a "ruler" (or more correctly, a scale). A scale is divided (commonly) into marks on each 1/32th of an inch, that is, there is a mark at every 0.03125 inch. There would also be marks noting 1/16th of an inch (0.0625 inch), the 1/8th inch (0.125 inch), ¼ inch (0.25 inch), and ½ inch (0.50 inch) spacings. Thus, our designs could "shoot for" numbers like:

- 3.00
- 3.50
- 3.625 (3.62)
- 3.6875 (3.68)
- 3.71875 (3.72)

Note that these numbers get more "uneven" in the sense that they are increasing by some smaller fractional amount (1/2, 1/8, 1/16, and 1/32). Also note the "rounding" of the numbers to *even* numbers. Why choose an even number? Well, if the overall part was chosen to be 3.68 inch, if one wanted to place a hole half way across the part, that hole would be placed at 1.84 inch. If the overall part was chosen to be 3.69 inch, then the hole that is half way across would be at 1.845 inch, but this 3-place dimension "implies" a "tighter" tolerance than is actually required (more on this later in the section).

Also note that this scale with the fractional markings actually had a scale (on the other side) that had graduations in 0.020 inch increments. So, parts could have been designed in increments of "tenths," such as 3.120 inch, but again, fractions were more commonplace.

This brings up a discussion on *nominal* or "evenness." Designers are more apt to choose even numbers, and by that, given the choice between:

- 3.00
- 3.10
- 3.25
- 3.26
- 3.27

I would say that a designer would choose them in the order above (top to bottom), without consideration to size, *just* by the numbers. I believe that 3.25 could be chosen over 3.26 because 3.25 is 3 ¼. 3.26 does have an advantage in that it is an even number (easily divided in two). 3.27 "suffers" in that it is not a common fraction and an odd number. Above discussion assumes that one wants the numbers at two decimal places.

4.8.2   *United States Engineering Units Versus International System of Units*

My comments about dimensioning/tolerances are useful in either system of units (inches or millimeters). In the United States, we may start a design thinking that 3.000 inches is nominal or "the place to start." In Europe, the "same" place to start might be 75 millimeters (which is equal to 2.953 inches).

If a design starts in the United States at 3.000 inches, those drawings are exactly *converted* to 3 × 25.4 = 76.2 millimeters if the product is to be manufactured in Europe. I'm trying to make a distinction between "conversion factor" (inches to millimeters) and "designer origin mindset." So, if I was a designer in the United States (with a US education), who was designing a part for a European firm, I would probably create a design that had its start with an "even" millimeter nominal dimension, so I would start with 75 millimeters as that size (instead of choosing 3.000 inch). My choices in that general range of numbers would be among:

- 70 millimeters
- 75 millimeters
- 80 millimeters

4.8.3 The World Before and After CNC-Controlled Machine Tools

With machine tools before the computer was added to them, a machinist would "dial in" by hand a reading on a dial and then "make the cut." For example, on a lathe turning machine, if a 2.000 inch diameter rod (2.000 nominal, with a tolerance of +/−0.002 inch) was to be the "final" diameter, an initial cut would be made on the rod; a measurement taken (physically) of the part might be, say, 2.015 inch diameter (initial cut would always be made so that it resulted in the part being slightly larger, in a sense, "sneaking-up" on the wanted dimension). The tool on the lathe would then be "zeroed" on the part diameter, and then the dial would be moved to 0.0075. A cut, 0.0075 deep on the part, would theoretically take that part down to 2.000 inches. If another measurement were to be taken on the part, it would likely measure close to 2.000 inches diameter. Of course, the complication of adding knurling, plating, or paint is another factor to consider, but let's keep it simple for now and not consider other finishing operations. After the part was removed from the lathe, it would be inspected. If the part was between 1.998 and 2.002 inches, it would be "passed" as correct and meeting specification. This general procedure, of zeroing, dialing, cutting, and inspecting, produced "industry acceptable" standards for attaining parts (within a tolerance of +/−0.002 inch) in a productive manner. That is, the machines, machinists, and inspection processes provided parts in a "reasonable manner." If the tolerances were made tighter, say, +/−0.001 inch, the parts would be more expensive due to either of the machine, machinist, or inspection process being more difficult, resulting in a slower time to produce acceptable parts. Also, asking for parts with looser tolerances, say +/−0.003 inch would *not necessarily* result in less expensive parts, as +/−0.002 inch would be "standard practice." Of course, machine shops would enjoy looser tolerances as their rejection rates would go down, but these savings were likely not passed on to the customer. Again, I'm making an argument here as to specify tolerances in cooperation with your suppliers to find a balance between what the part needs and what can be reasonably (or typically) produced.

Note that (see Ref. [14], Machinery's Handbook) a turning machining operation, under normal conditions, would produce work in the range of Grades 7–13. For a 2.000 inch diameter, a tolerance of +/−0.002 inch is reasonable (but difficult). From Ref. [9], "normal consistent accuracy" would be +/−0.005 inch.

Now, with CNC-controlled machines, part costs have been reduced. This is due to several factors:

1. Total time is reduced to produce part. Once the machine is programmed, the machine will move according to that program, basically eliminating the "machinist." So, no "dial moving" is needed. (Of course, a machinist is still needed to "oversee" the operation in a general sense.)
2. Less inspection is required as the machine has eliminated "operator error." Inspection is still needed of course.
3. The "machine" can change tools, speeds/feeds, axis of machining, and just about any machining operation in a machine-controlled manner. This is faster than can be done with a human operator.

4. Small changes to the part design are very easy to accomplish – a quick revision to the program, and the new revision parts are quickly produced. Even extensive changes still only mean relatively simple program updates.
5. Machining "factors" can be built into programs to *adjust* for any machining situation that might occur. For example, if a punch results in a hole being out of tolerance in an x-y coordinate, the program can be "adjusted" to bring that hole position into tolerance.

Flat metal punching by a CNC-controlled machine (a "strippet") allows x-y table movements and tool changes via head rotation. This allows reduced set-up time, elimination of fixturing, and machine changes at tremendous time savings over the previous generation of machine tools.

As input to a CNC-controlled machine is a computer program, parts are the result of only digital data that is transferred to those machining programs. This allows extremely quick (and "error-free") transfer of "design intent" to "finished part."

4.8.4 Overall Size and the Design

To continue the general topic of dimensioning and how it relates to design, Some constraints will usually be the start of a design. Here are some examples:

The general assumptions for the examples are:

Minimum size and weight, Cost (Chpt4), are required in the design.

Minimum clearance between one object and another is 0.010 inch. This obviously varies in a real design and depends on the objects and environment that the enclosure will be used.

Nominal thickness of enclosure is 0.050 ("thickness tolerance" will be discussed separately), but the thickness will simply be at 0.050 for now.

State-of-the-art designs may require everything from custom sheet metal gauges to "fixtured" solutions that include "go/no-go gauges" that essentially allow parts to be constructed with *no* (or,*very* little) tolerances. The examples will not consider these types of situations but, rather, more "normal" design circumstances. Again, given an unlimited amount of time and money, just about any exception can be taken with the examples. Designs usually proceed with one of the following two types of thinking: See Fig. 4.2.

"Size-goaled" Design: An enclosure must be smaller than 3.5 inches. This is because the previous product was 4.0 inches (or the competing product is at 3.6 inches). In this case, we start with an overall enclosure size of 3.5 inches. If a single component that is 3.0 inches long is to be put into our enclosure, we immediately know that the "nominal" distance between the inside of the wall and the component will be: 3.5 minus 3.0 minus (2 times .050) divided by 2. This equals 0.2 inch. We have "accepted" the 3.5 inch outside dimension and are "accepting" the 0.2 inch clearance between wall and component.

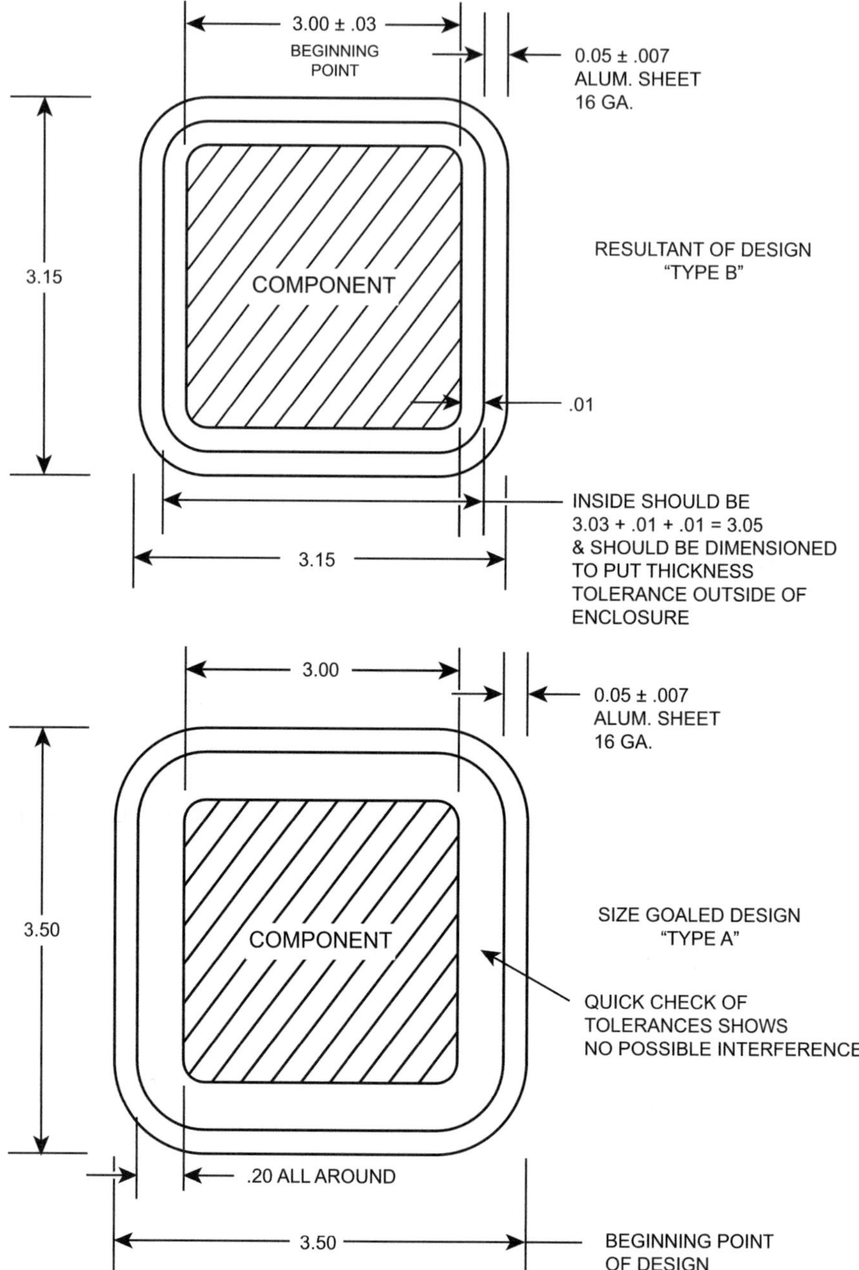

Fig. 4.2 Overall size and the design given

"Resultant of Given" Design: A single component that is 3.0 inches long is to be put into an enclosure. That "sets" the (minimum) size of the enclosure at 3.0 + clearance per side (0.010, estimated for now) + thickness of both walls of the enclosure (in just the "x" direction); thus the overall size of the enclosure is:

$$3.0 + 0.010 + 0.010 + 0.050 + 0.050 = 3.120 \text{ inches}$$

However, we haven't stated the tolerance on the "3.0 inch component." Let's say that this component's *maximum* size is actually 3.03 inches. That takes the minimum size of the outside of the enclosure to 3.12 + 0.03 = 3.15 inches.

Let's take a look (more closely) at both the clearance and thickness numbers and investigate what the tolerance actually is on those numbers.

There will actually be a tolerance on the thickness of the enclosure. Let's investigate this for various materials:

Sheet metal thickness: Sheet metal is normally available in standard gauges that have standard tolerances on thickness. 16 gauge aluminum sheet, which is "nominally" 0.050 inches thick, has a +/−0.007 inch tolerance on its thickness (5050 alum, 100 inch stock width) per [10]. It's been my experience that the thickness will "always" come in on the "thin side" of the tolerance (but could come in on the "thick side"). General deburring of a plate (before punching) will take 0.001–0.002 inches off of the thickness, while some plating will add back the 0.001–0.002 inches. The designer should plan on a nominal gauge thickness (for most designs) unless it's actually warranted taking a specific thickness tolerance into consideration. Note that clearances should be checked at both their maximum and minimum conditions to ensure proper fit both in assembly and customer use.

So, designs usually proceed as either the Size-goaled or "Resultant of Givens" type of design. The Size-goaled design may result in some saving of time, however, the Resultant of Givens design usually is a better methodology for getting to the *minimum-sized design*.

4.8.5 Theory of Tolerancing: The Need

Tolerancing on part dimensions is needed as parts cannot be produced *perfectly*. Manufacturing techniques do not produce perfect parts. This is probably obvious. The actual amount of tolerance is based on a few (probably competing) factors:

1. Cost (larger tolerances are less expensive to manufacture).
2. Like parts (the same part) need to be *interchangeable* – all parts made to the tolerances must work.
3. Parts that mate with other parts must do that with all parts at the extremes of their tolerances.

So, the "default procedure" for specifying a tolerance is to:

1. Choose the tolerance for the reasonable or most common manufacturing process. For example, if the part is a 0.25 inch diameter part, say 2.00 inches long, a common way to produce this part would be with a CNC-controlled spindle axis machine (or a lathe). The "industry-accepted" tolerance on the diameter might be + or −0.002 inches. If the part was to be an *investment casting*, the tolerance on the diameter might be + or −0.001 inches.
2. If the above is acceptable to the design, look to increase the tolerance even more, checking back to acceptability in the overall design. Increasing the tolerance will allow more parts to pass inspection, which should (could) result in overall cost reduction.
3. Tighten the tolerance even though the most common manufacturing process will not produce that tolerance *if the design dictates that tighter tolerance*. Check with the part manufacturer if that tighter tolerance will be achievable (reasonably) or at what cost.

Each dimension (location, hole size, angle, etc.) must have a tolerance, either explicitly stated on the drawing or as being a part of an overall notation on the drawing.

Some designs require parts with *very* low tolerances. That is, if parts are manufactured that exceed those tolerances, they will not work!

Most designs should allow "standard industry" tolerances to be cost-effective. These "standard industry" tolerances are found by:

- Consulting with part manufacturers
- Researching standards available in various industries
- Using previously attained knowledge by the designer (i.e., experience)

Tolerances are listed on drawings (documentation) in several manners. An individual dimension on a drawing may carry its own tolerance; there may be notes that define the tolerance on several dimensions or there be a note on the drawing that covers every dimension. In any case, *every* dimension does indeed have a tolerance (that should be specified).

There are several methods of dimensioning. For example, a dimension that is nominally 2.000, with a tolerance of +/−0.005, can be specified on the drawing as:

- 2.000 +/− 0.005
- 2.005/1.995
- 2.005 + 0.000 − 0.010 (unilateral dimensioning)
- 1.995 + 0.010 − 0.000 (unilateral dimensioning)

If the tolerance was an odd number (say, 0.003 *total*), this will produce an example of *bilateral* dimensioning, as

- 2.000 + 0.002 − 0.001

But, this would be a rather rare occurrence.

Unilateral dimensioning is advantageous when a critical size is approached as material is removed during manufacture, as in the case of close-fitting holes and shafts. Thus, for a shaft (where material is removed), the high side of the limit is shown first as:

- 2.005 + 0.000 – 0.010

However, it would be much more common to list this dimension as

- 2.000 +/– 0.005

The number of decimal places for the tolerance should match the number of decimal places shown for the basic dimension. For example, a dimension given as 2.000 would have a tolerance listed as +/−0.010 (not +/−0.01).

4.8.6 Theory of Tolerancing: Accumulation of Tolerances

Every individual part has overall dimensions (length/width/thickness) as a minimum. Parts usually have, in addition, features (cuts, holes, etc.) that need to be dimensioned. In dimensioning, it is very important to consider the effect of one tolerance on another. When the location of a surface in a given direction is affected by more than one tolerance figure, the tolerances are *cumulative*. For example, in Fig. 4.3, the location of Surface A can be controlled from either Surface B or Surface L:

In the "cumulative tolerances," the tolerance of the location of Surface A from Surface L is 2.000 + 0.000 – 0.010.
In the "base line dimensioning," the tolerance of the location of Surface A from Surface L is 2.000 + 0.000 – 0.005.

Thus, if the feature (Surface A) *needs* to be held more closely to Surface L, then that feature should be *directly dimensioned* from that same feature.

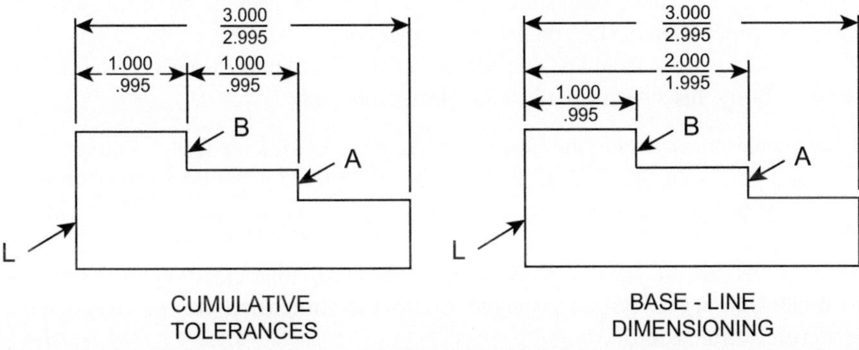

Fig. 4.3 Accumulation of tolerances

Every dimension should be analyzed to determine exactly what feature it should be controlled from (dimensioned from) so that tolerances don't accumulate from "irrelevant" features.

4.8.7   Inspection Dimensions (Critical Dimensions)

Parts can have hundreds of dimensions. To fully specify every feature on some parts takes a fair amount of time. Historically, every feature on a drawing was dimensioned to enable that feature to be inspected and found either in compliance with the specification (drawing) or non-compliant. With the advent of 3D CAD systems for designing and creating drawing documentation, it is completely possible to transfer all of the part information digitally, without any dimensions being specified actually on the drawing, and have that part manufactured at the "nominal specified dimension." That is, features that are 3.000 inches apart are "drawn" exactly 3.000 inches apart, and that number is an integral part of a file that the part manufacturer gets. Anyone who brings up that part file on their CAD system can query that file and see those features need to be 3.000 inches apart (with a stated tolerance as a part of the CAD file). In fact, inspection departments, if they had access to the CAD file, could inspect the part (perhaps with a 3D computer-controlled inspection system) and determine whether that dimension (3.000) was within tolerance.

So, the question becomes who exactly needs to know the dimensions and to what purpose is it that the dimensions are shown on the drawing documentation? The answer to that question is probably different for each organization. Some organizations have settled on not specifying every dimension but rather a subset of the dimensions.

It has become common practice to label dimensions on a part drawing those dimensions that determine whether the part has "passed inspection." On parts with only a (relative) few dimensions, that might mean that *all* of the dimensions are considered "critical." On parts with hundreds of dimensions, a subset of those dimensions could be considered "critical" and must be verified by standard inspection procedures. Those procedures could be 100% inspection or some agreed-to sampling process that is less than 100% inspection.

What dimensions would be considered "critical" and thus inspected? Candidates for specifying this subset (of all of the dimensions) are:

1. Is the feature critical to the function of the part? If a feature (holes, cutouts, etc.) mates to another part, then this may be considered a "critical dimension." If somehow *safety* is involved, then that would place the dimension in the "critical" category of *must be inspected.*
2. Does the feature have a "special" (nonstandard) tolerance which places that dimension in the "critical" category? If so, that dimension may be considered a "critical dimension."

3. Have certain dimensions (when parts have been prototyped) been problematic in that they seem to cause the part manufacturer some problem which causes an issue to develop with that part? If so, that dimension is another candidate to be labeled a "critical dimension."

Some issues arise with this concept of "critical dimension" in that *any* feature (no matter how "insignificant," such as an internal radius), if not to tolerance, can cause an issue. It's just that many of the dimensions have a high probability on *not* causing any issue, and thus, only a small percentage can be considered "critical."

4.8.8 *True Position Dimensioning*

True position dimensioning, or more specifically, geometric dimensioning and tolerancing, is a means of specifying engineering design and drawing requirements with respect to actual "function" and "relationship" of part features. Furthermore, it is a technique which, when properly applied, ensures the most economical and effective production of these features. The major objective of the system is uniform interpretation among design, production, and inspection groups. I'm not going to present the details here but rather a brief overview of the basis and reasoning behind geometric dimensioning and tolerancing (GD&T). The current authoritative document governing the use of "GD&T" is shown in Ref. [15–17].

One of the basic principles of GD&T is the idea changing from a "rectangular tolerance zone" to a "circular tolerance zone." Normal positional tolerances in a rectangular coordinate system permit a "square" tolerance zone, while the tolerance zone changes to a circle with true positioning (with the area of the circle being larger than the square by 57%).

The manufacturing industry is turning to geometric tolerancing primarily because it increases productivity and reduces cost. The advantages include:

- Increased productivity by specifying maximum but workable tolerances, in many cases permitting manufacturing variation beyond the tolerances specified by dimensional tolerancing (the older system of tolerancing)
- Interchangeability of mating parts by specifically stating design requirements in relation to the function of the part and making possible the use of time saving functional gages
- Uniformity in drawings and in their interpretation, ensuring effective communication between engineering, manufacturing, and quality control

If a pattern of four holes is located from part edges and a distance is given between holes (including the tolerance on the four dimensions), it increases the chances of mating parts that won't fit together. Instead, if the pattern of four holes is located from part edges and *basic* (exact, theoretical, untoleranced) *dimensions* are used for X and Y between holes, along with the addition of a *"true position"* dimension, there will be no ambiguity as to the pattern's potential positions.

If a single hole location with +/−0.005 inch tolerances given with dimensional tolerancing is replaced with a positional tolerance zone of 0.014 inch *diameter*, the tolerance "area" grows from a rectangular (0.010 × 0.010) area to a circular $\pi/4$ $(0.014)^2$ area, with a resulting 57% increase of area. This can equate to "real-world" rejection rates going from four parts per 100, being reduced to two parts per 100. Thus, there are more acceptable parts just by changing from dimensional tolerancing to positional tolerancing.

Complete true position dimensioning consists of:

1. Basic locating dimensions (where tolerances do not apply)
2. An expression (notation) such as located at true position within.XXX DIA (or located within.XXX R of true position) added to the feature to be located
3. Datum references

To prevent misunderstanding, true position should always be established with respect to a datum.

Actually, the "circular tolerance zone" is equal to the positional tolerance and the axis of the feature (usually a hole) must be within the cylinder. The center line of the hole may coincide with the center line of the cylindrical tolerance zone, it may be parallel to it but displaced so as to remain within the tolerance cylinder, or it may be inclined while remaining within the tolerance cylinder. Thus, the axis of the holes are not only defined in 2D but rather in 3D, which is a much more powerful description of acceptability for a hole pattern. This is illustrated in Fig. 4.4.

Methods of indicating geometric tolerances, by means of GD&T, define conditions of straightness, flatness, parallelism, perpendicularity, angularity, symmetry, concentricity, and roundness. For example, a note attached to a cylinder that states *"straight within* 0.010 *total"* is a "shorthand" for the statement:

Regardless of the actual size of the feature, any longitudinal element of the surface must lie between two parallel lines (0.010 apart) where the two lines and the nominal axis of the feature share a common plane.

Thus, a "standardized" note conveys very clear meaning universally understood by the designer, the fabricator, and the inspection departments.

Datum features are powerful tools in GD&T. They should be selected based on their ability to be:

- Functional
- Mating surfaces
- Readily accessible
- Repeatable

Some other facts about datums:

- A datum is a theoretically exact geometric reference.
- Primary datums provide feature orientation.
- Datums are assumed to exist on the part itself.
- Datums are established from tooling points or surfaces.
- Surface plates and V-blocks are simulated datums.

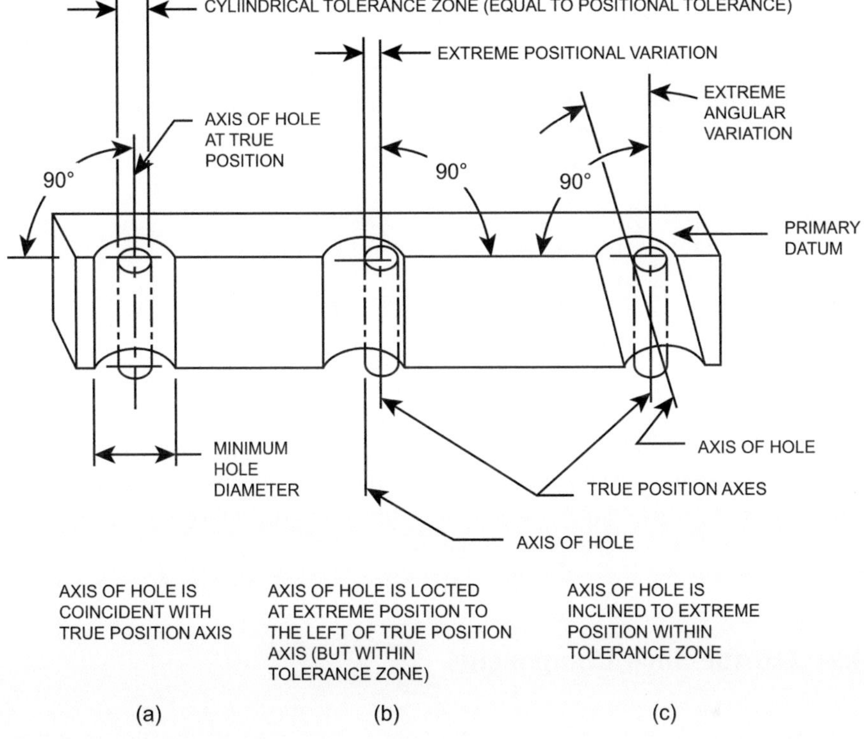

Fig. 4.4 Positional tolerance controls perpendicularity

- Datum features (which are actual features of a part mated to simulated datum surfaces) contain inaccuracies.
- A datum reference frame consists of three mutually perpendicular planes.

GD&T also has the concept of "modifiers." Modifiers are used in conjunction with feature control symbols to identify how the geometric tolerance zone is affected by the size tolerance. The modifiers that are used most commonly are maximum material condition (MMC) or regardless of feature size (RFS).

MMC is shown on a drawing as a "circled M," while RFS is shown on a drawing as a "circled S."

MMC indicates the most material possible in an object and is used in a feature control symbol. It means that the geometric tolerance will only affect the size of the part at MMC. (The true position symbol "circle with cross hair" implies MMC).

RFS means that the geometric tolerance will affect any produced size of any object. Most geometric characteristics (straightness, flatness, etc.) imply RFS.

Let's look at how the use of modifiers affects a geometric tolerance zone. We'll look at three examples:

- Conventional (non GD&T)
- MMC
- RFS

A bar with the dimensions 0.25 +/− 0.02 will all be produced in a range of acceptable sizes from 0.27 to 0.23.

With conventional tolerances, there is *no control* on the *straightness*.

With GD&T, a straightness symbol, and MMC (or RFS), the maximum out of straightness varies as the produced size varies:

Size	Max. out of straightness MMC	Max. out of straightness RFS
0.27	0.03	0.03
0.26	0.04	0.03
0.25	0.05	0.03
0.24	0.06	0.03
0.23	0.07	0.03

Again, geometric tolerancing is a powerful technique to help the designer think about what are the "control features" of the part design and how the part will actually be inspected to determine whether the features are in compliance with the drawing (design). Additional references are listed.

4.9   Off-the-Shelf Components

The designer of electronic enclosures must always start the design with a search of what has already been designed that could be utilized in the new design. The search should include inside (within the designer's own company) sources or OEMs external to the company.

Examples that are already available to the designer are:

- Fasteners, washers, standoffs, threaded inserts, pins, springs
- Heat sinks, insulators
- Cabinets, racks, slides
- Rubber grommets, bumpers, shock isolators, gaskets, O-rings
- LEDs, display products, switches, pushbuttons
- Cables, connectors
- Labels, nameplates

Sometimes the off-the-shelf part is not quite what is needed. The decision to either make the new part (exactly what is wanted), contact the OEM to see whether the slightly different version can be attained, or adapt the design to the existing part needs to be analyzed. The decision should converge on the normal constraints of cost, time, and conformance to specification. Also, the catalogs of the OEM suppliers usually contain relevant design information that can be used to design similar parts to those within the catalogs.

Consideration should be given to the reuse of parts from another design. For example, if a new slimmer enclosure is needed, can the upper part of the present enclosure be reused?

4.10 Prototyping

I've tried to write this book so that it will be relevant to designers going forward in time, that is, to not put technical material in the text that quickly would become obsolete. This section on prototyping is "at risk" in that I've witnessed that the techniques used for prototyping have undergone the biggest changes in my 30+ years of design engineering, and I can see them continuing to change at a rapid pace, making a lot of what I now say about the particular processes obsolete (perhaps before publication). However, one thing that will not become obsolete is the *need* for prototyping.

Prototyping is essential to the efficient design of any engineering product. Producing a quick, relatively realistic representation of the "final idea", gives everyone on the team a sense in what direction the project is moving. The prototype will either move the project forward or cause a reassessment of the present thinking. The prototype can be of any portion of the entire project, from just a small portion of the product to the entire finished product.

I've seen many instances of (very) crude prototypes providing amazingly quick information to the design team. Paper, cardboard, and scissors can quickly illustrate some aspects of a design – again, the speed of prototype production cannot be overemphasized.

Drawings, by themselves, do not give everyone on the design team an idea of what the design "will be." Even three-dimensional representations will not completely convey the design. There are some design engineers who can just look at drawings and see that the final design will achieve the objectives; however, there are some people who do not have the ability to interpret those "lines and numbers" into something they can imagine and that they need to touch and feel an object to determine whether it "will work." *size* is one aspect that seems to be more easily conveyed by a physical prototype (than on a drawing). Certainly, any design feature that has human "interaction" will need to be actually picked up and used by the design team to understand the design feature's ability to achieve the design's objectives. Just about anything that the customer's hand will touch needs to be modeled to see how that "mind-hand interaction" will work. An entire discipline, ergonomics, is dedicated to determine those best decisions. For example, if a door is to occur on a product, some common questions occur (where a prototype will help answer the questions):

- What mechanism will open the door (and close it)?
- How is the door repaired once broken (under abnormal use)?
- Will the door open side to side or up and down?
- How well does the door need to seal (once it is shut)?
- How much force is needed to open the door (and close it)?
- Does the door need to stay open (without holding it open)?
- Does the door need to be locked?
- Does the door slide, hinge, roll, or otherwise move to open?
- What (on the door) serves to open and close the door?

Speed is a key factor in the production of a prototype. The prototype is purposed to provide "quick information" on the design itself, so finding out whether to pursue a design direction, or whether to change direction, is paramount to the process. Also, a determination on how well the prototype will portray the design intent must be made. That the prototype is so "crude" that it allows a design direction to continue just because it didn't accurately mimic the final design enough will not help the overall process. A balance must be made in determining "accuracy" vs. timeliness in the prototype process. Once "speed" is determined to be the *key* factor, cost must then be secondary. That is, costly prototype production that saves overall product delivery parameters (time and conformance to specification) is a solid investment.

In-house prototyping capability vs. subcontracting out the prototyping is always a question to be answered (by the project management team). Some companies feature having the prototyping capability in-house, as this has the advantages of:

- Scheduling of the prototype (against competing needs) is better controlled (should result in quicker prototypes being produced).
- Profit in the prototyping process itself is kept in-house.

Having the prototyping process being subcontracted out may be better due to:

- Latest processes are more available (usually) out of house. This is due to the rapid pace of change in prototyping processes and the capital cost of equipment needed for these prototyping processes.
- No maintenance or instruction is required on prototyping equipment.

The decision (in-house or contracted prototyping) must be analyzed as time moves forward. For example, the cost of 3D printing (of plastic parts) in-house was generally prohibited as capital cost, size of machine, training required, and pace of technology change were all too high before the year 2000. But, by 2005, having a 3D printer right there in the office became quite common.

Getting sheet metal or metal parts prototyped can also be done by various 3D printing methods but, at present, continue to be in the domain of "quick-turn" prototyping facilities (utilizing the standard metal fabrication methods of drilling/machining/punching).

Another consideration of the prototyping process is whether the prototypes are fabricated at either:

A. The same vendor that will manufacture the production parts
B. Vendor that is independent of the future production part location choice

An advantage of utilizing the "same vendor" is that some of the issues are worked out with the prototype parts that will "smooth the flow" of production parts. However, sometimes the speed gained by using a vendor that is independent of the future production parts will get the prototype parts in hand sooner. Again, no one answer will fit all unique situations.

Appendix

This appendix is provided to the EPE designer to make them aware of some common sheet metal punching and forming methods. Also, there is some information on the "best practices" of dimensioning sheet metal designs. The EPE designer will be designing quite a few parts that are made out of sheet metal as sheet metal provides the strength that many individual parts of enclosures will need. Common metals used will include aluminum and stainless steel.

Much of the material shown here has been commonly shown in corporation design guides and the EPE designer will hopefully find useful information. As usual, the designer should also consult with sheet metal fabricators to see if this information is consistent with their fabrication process.

Dimensioning of Sheet Metal on Drawings

Dimensions should be placed to prevent having to add or subtract material thicknesses and tolerances, and should be supplied according to the *function* of the part. This is illustrated in Fig. 4.5 for a base and base cover. Note here that normally, all dimensions should apply to the outside of the part. When inside dimensions must be utilized to assure fit of mating parts, the word "inside" can be added to the dimension to emphasize the idea that the part's dimension "mates" with another part (and is therefore critical). The most important aspect of this is that the dimension is on the item that is the most critical, and that is that the cover will, indeed, fit on the base.

Dimensioning Bends

The dimensioning of bends in sheet metal parts should be applied from a tangent or extension point and not to the radius center as shown in Fig. 4.6. Tangent or extension points are better because these points can be *measured*, and this is the

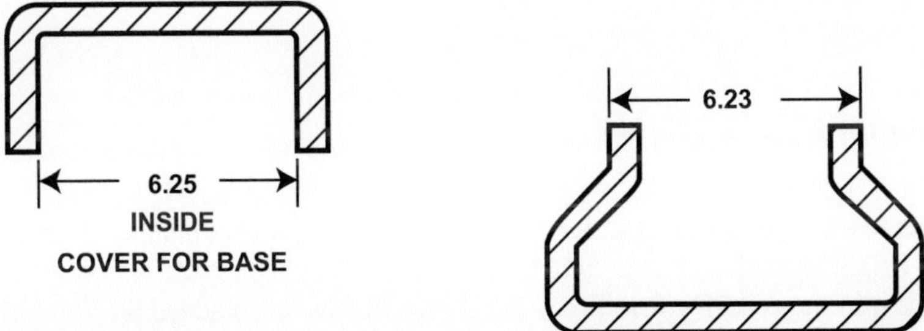

Fig. 4.5 Feature control

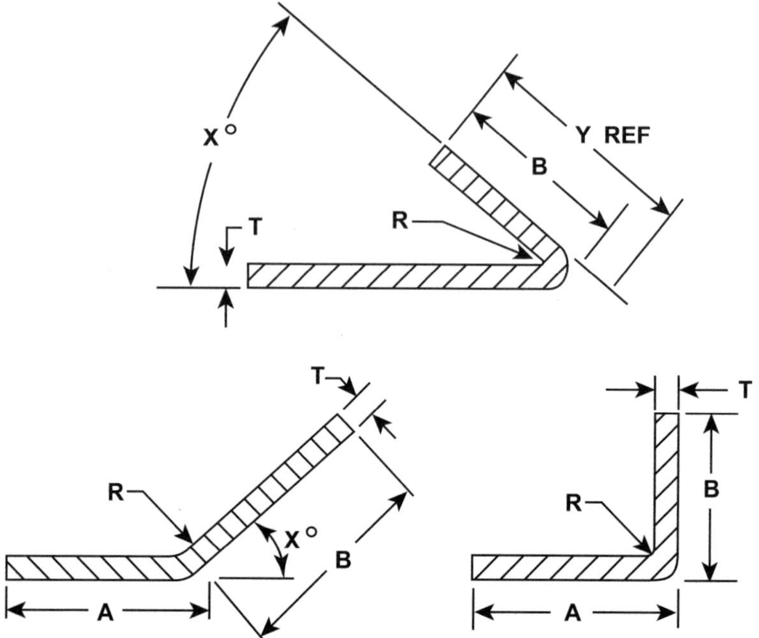

Fig. 4.6 Dimensioning bends

important goal. Anytime a dimension is given to a point which cannot be measured (easily), ambiguity can result. Another point to be made is that, usually, the center of a radius is not what actually needs to be located in the design, but the part edge is the more critical. A more simple "rule-of-thumb" may be that if the center of the radius is not required to construct the part, then the dimension to the center of the radius is probably not required.

Note also in Fig. 4.6 that the dimension "B" in the acute and obtuse angles could be replaced by a dimension from the base to the upper height of the angle (as the right angle is dimensioned). This dimension would be even more valuable as it is easily arrived at (inspected) with a height gage.

Minimum Flange Height

In order to provide sufficient material for the proper forming of bends, the minimum flange height should be as shown in Fig. 4.7. If the design calls out a dimension less than the minimum 2.5T + R in Fig. 4.7, shock will have to be added for bending and then cut off, requiring an additional operation at added cost.

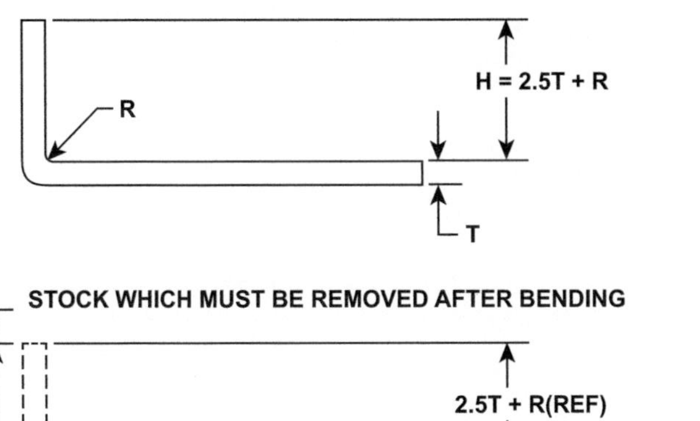

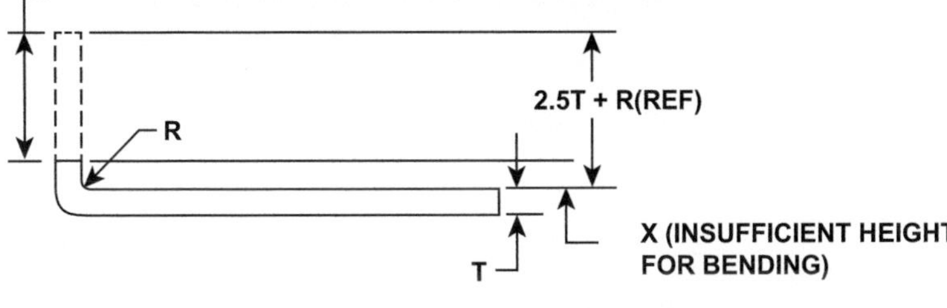

Fig. 4.7 Minimum flange height

Minimum Distance Between Bends

Parts having "Z" bends similar to those shown in Fig. 4.8 should have the minimum distance between bends no less than the values shown. Also, there is a minimum aspect ratio between bends in a "U" shaped piece. As shown, a forming tool cannot be made small enough to avoid interfering with metal on a bend already created. It is a good practice to imagine (with a sketch on a drawing) the forming tool *itself* on a relatively high depth-to-width ratio part to see if the forming tool is practical in size.

Location of Holes Adjacent to Bends

In order to prevent distortion of a hole pierced or punched before bending, the minimum distance between the edge of the hole and the edge of the bend should be as shown in Fig. 4.9. This distortion takes the form of an elliptical hole with an increase in material thickness which would be unacceptable in most design efforts. Note, calculate "X" dimension so that "Y" is not less than 1.5T + R.

For slots which lie parallel to the bend as shown in the figure, the following minimum distances shall apply:

When L = up to 1 inch,   "A" = 2T + R
When L = 1 to 2 inches,   "A" = 2.5 T + R
When L = 2 inch or greater,   "A" = 3T + R

MATERIAL THICKNESS	LENGTH "A"
.000 - .040	.188
.041 - .080	.500
.081 - .125	1.250

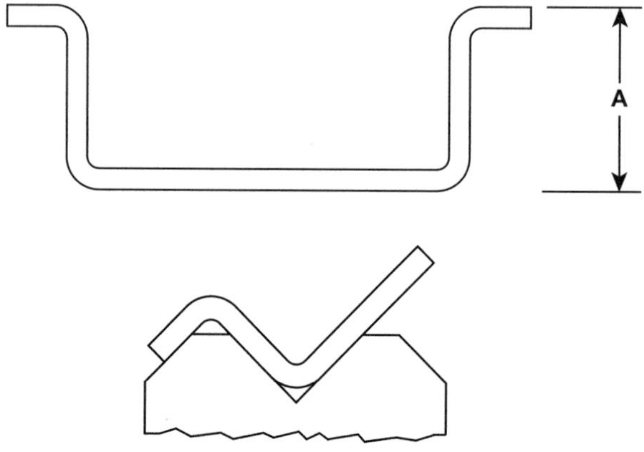

Fig. 4.8 Minimum distance between bends

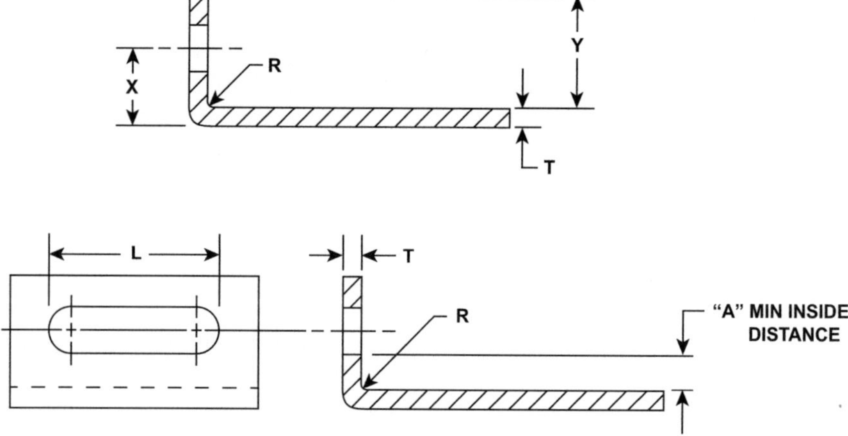

Fig. 4.9 Location of holes adjacent to bends

Location of Holes to Prevent Distortion

To prevent distortion of the edge of a part or of the material between holes, minimum distance calculation for punched holes must consider:

A. Material used, its thickness, and physical properties
B. Hole shape and size
C. Hole application

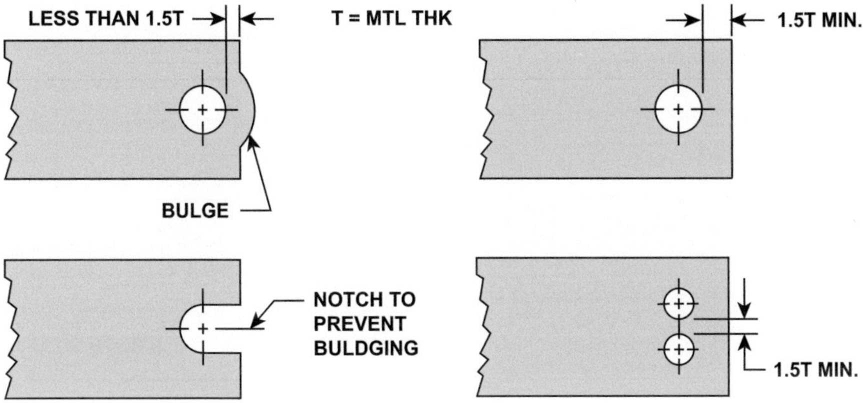

Fig. 4.10 Location of holes to prevent distortion

Nominal minimum distances are shown in Fig. 4.10. The variables listed above should be used to determine special requirements. For example, edge distortion may be avoided by notching, as shown in the figure.

Punching Holes

Round holes may be economically punched in sheet metal parts provided the diameter of the hole is larger than the thickness of the material. If the hole diameter is less than the material thickness or less than 0.032 inch, the hole will have to be drilled.

Normal punching tolerances are +/−0.010 inch from edges and +/−0.005 between holes. These would be considered normal for numerical controlled (NC) machines such as those made by Strippet and Amada. More on this in Sect. 4.8 on tolerances.

Figure 4.11 illustrates two methods used to dimension holes. Only one datum in each direction should be used. Use the lower left or upper left corner of the part for datum unless there is a good reason to do else wise.

For a "cut" corner (flat panels, plates, etc.), use the corner as datum references. For folded corners, the first hole should be the datum reference (tolerance +/−0.03 preferred).

Typical punched slot dimensioning is shown also in Fig. 4.11. The drawing would specify basic sizes, detail dimensions, and tolerances required. In this case, a +/−0.010 tolerance was entered in the drawing title block.

Standard punches exist at most sheet metal vendors. They should be able to provide you a listing of the standard punches they have, so that you can avoid a "custom-sized" punch.

Punches include variations of round, square, rectangular, and oblong punch sizes for use on aluminum and mild steel sheet. Inside and outside corner radius punches are also available in most common radii. Special punches such as standard connector cutouts may also be stocked at some vendors. Standard sizes and standard toler-

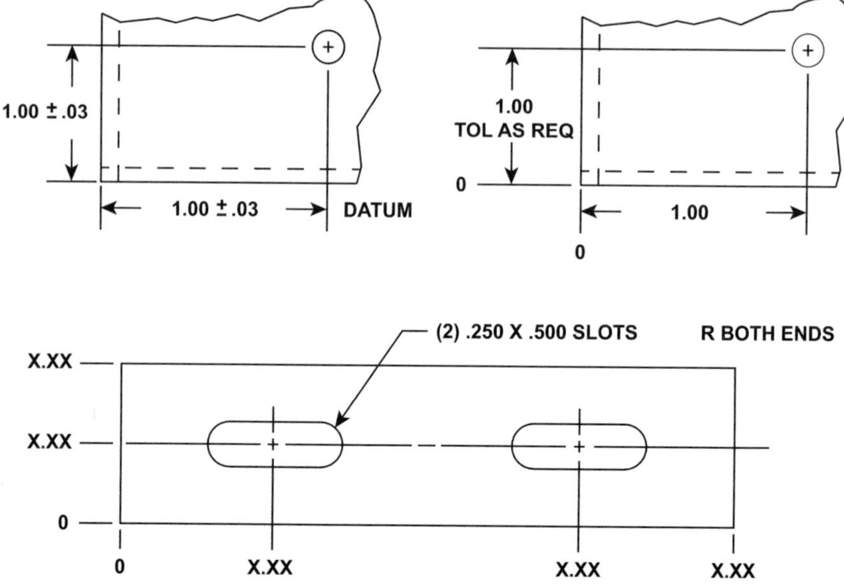

Fig. 4.11 Punching holes

ances should be used whenever possible, but in some cases, a special tool may be justified. In fact, even complete dies that blank entire parts are justified if the quantities of parts are high enough.

Bend Relief

The three designs illustrated in Fig. 4.12 are not desirable for quality or economy. In these examples, the metal at the base of the tab will tear in forming resulting in stress risers which could cause ultimate failure of the part.

Adding a cutout, which provides relief at the bends, is also shown. These examples represent good designs with relief for the bends to prevent tearing and to minimize fatigue under stress. The relief, which can have a radius to it (as shown) or can be "squared-off," is basically deep enough to be at the tangency point of the radius.

The relief is approximately a material thickness wide, but is generally not any narrower than 0.03 inch due to limitations on punch width. In general, this relief would not be dimensioned on the drawing; rather, it would be called-out as "Use Minimum Bend Reliefs" with the general drawing notes.

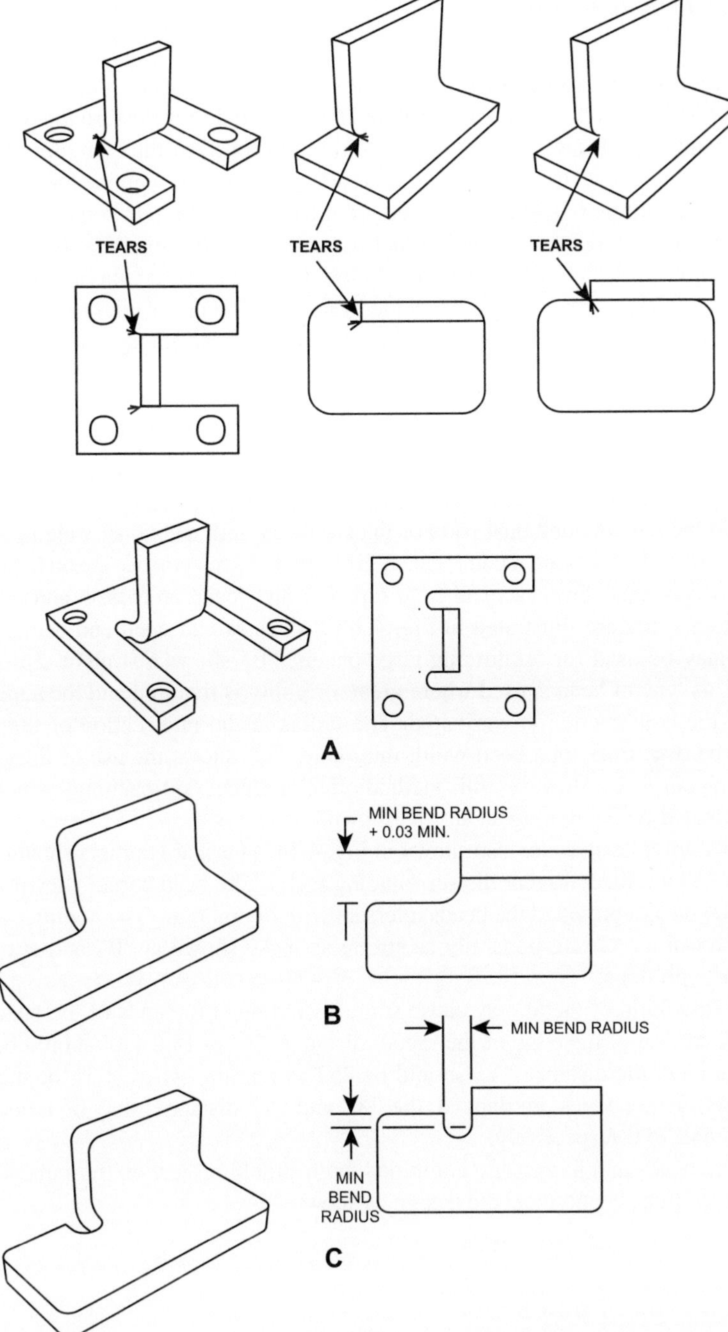

Fig. 4.12 Bend relief

Corner Construction

On all formed parts which involve adjacent flanges, relief cuts or notches should be incorporated to prevent tearing or wrinkling of the metal during the forming operation. The minimum allowances that must be made are illustrated by the examples shown in Fig. 4.13. Again, note that bend lines are at radius intersection with the flat surface.

An important note is to be made about corner construction. In some cases, extra strength may be required in a design that warrants the extension of both connecting flanges at the corners as illustrated by the closed corner construction of example "C' in Fig. 4.13. This corner would be welded (either inside or outside) to join the two perpendicular flanges to each other and greatly increase the strength of the member.

Beads and Gussets

To avoid the cost of additional parts or thicker metal with additional weight, formed beads (ribs) and gussets are recommended. Examples of such practice are on large areas of side panels, small brackets required to support heavy loads, on chassis, and cabinets.

Various types are illustrated in Fig. 4.14. "A" shows an open end center bead, which may be used for reinforcing large panels. "B" shows a straight closed end bead. This type of bead is used where greater rigidity is required and the additional cost of die construction is warranted. The radius at the intersection of the beads should be two times total bead width minimum. "C" shows the use of flanges for stiffening purposes. However, this method often produces "oil-canning" which may be eliminated by depressing an area as shown.

Bead corner designs are also shown in Fig. 4.14. The sharp corners shown in the bead at "A" are likely to tear the surrounding metal. The bead corner shown at "B" is a better design provided the beads intersect with liberal radii. The nonintersecting beads shown at "C" are generally as strong as those shown at "B" and are much cheaper to produce.

The stretching of metal may cause some wrinkling at the ends of the beads. To avoid distortion of the edge of the metal, distance "X" of Fig. 4.14 should be 40T minimum and the distance "Y" should be 25T minimum (where T is the material thickness). If the edges are flanged, the "X" and "Y" distances may be reduced to 30–35T and 15T, respectively.

Where beads and flanges are combined, both should project on the same side of the part for most economical die design as shown.

Minimum Bend Radii

There are recommended minimum bend radii for various materials and tempers that can be formed parallel to the grain of the material without cracking. Going below the recommended bend radius for a minimum bend should not be resorted to unless

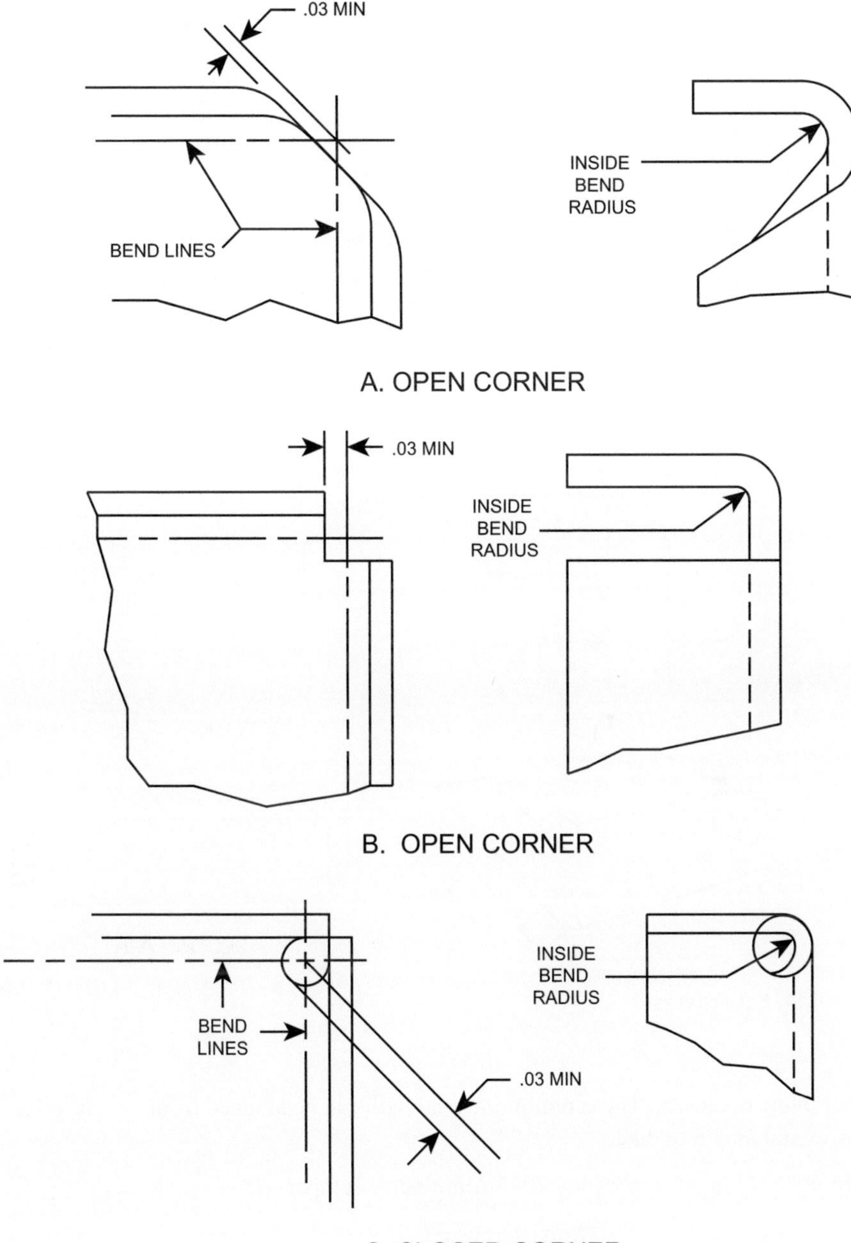

Fig. 4.13 Corner construction

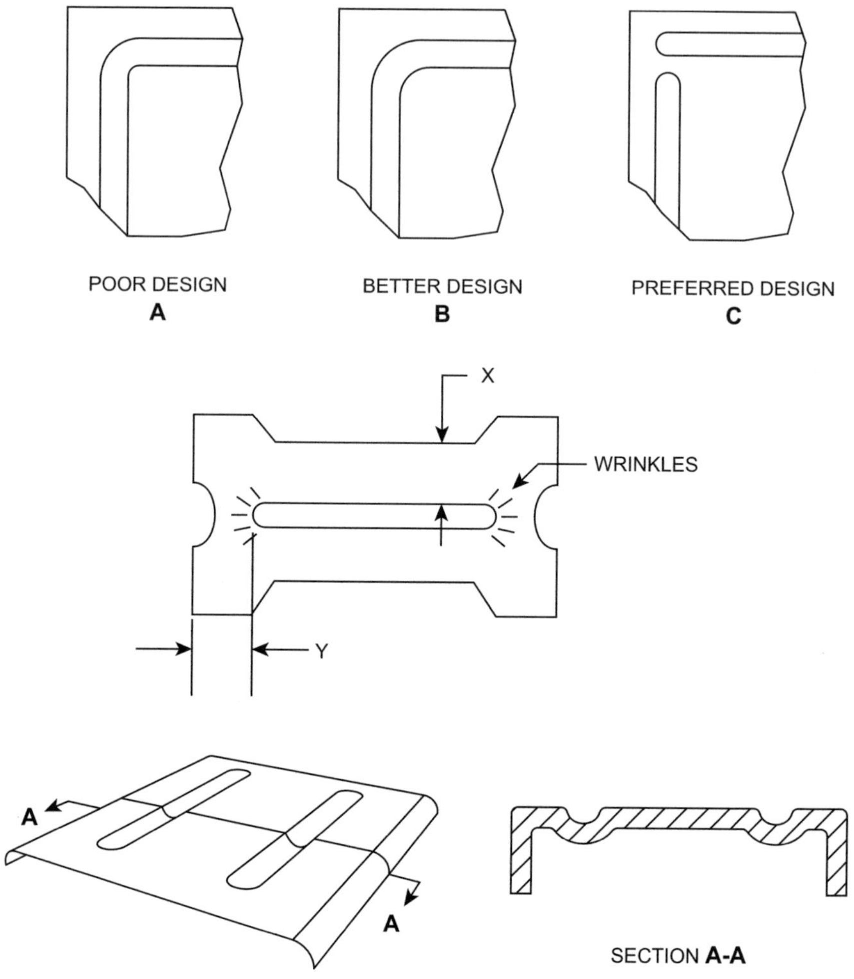

Fig. 4.14 Bead designs

absolutely necessary. These minimum bend radii can be attained from various references and may look like:

For 90° cold bending of Alloy 7075 Aluminum, Temper –T6,

Thickness 0.016   Min. Radius = 0.03–0.06
Thickness 0.032   Min. Radius = 0.09–0.16
Thickness 0.062   Min. Radius = 0.25–0.37

Normally, a note in the general note area of a drawing states: "Use Minimum Bend Radius (Unless Otherwise Specified)" which means to the sheet metal vendor

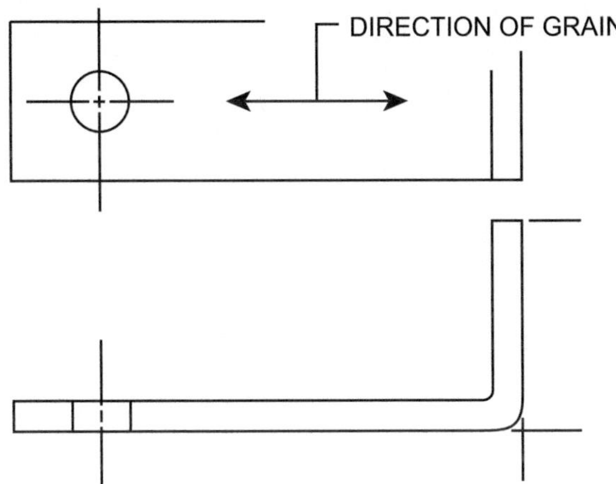

Fig. 4.15 Grain direction

that minimum recommended bend radii are to be used. Otherwise, a radius is called out on the drawing as preferred. Important to note is that bends should be as *generous as practical* to increase the overall strength of a part but are generally on the order of a material thickness.

Grain Direction

The grain direction of the material should not be specified on the drawing unless a bend radius is required that is smaller than the recommended minimum. In which case, the bend should be specified *across the grain*. This is illustrated in Fig. 4.15 showing that the bend direction is 90° to the direction of the material grain. Grain direction may be specified for minimum bend radii in hard or spring temper sheet or to show direction for a decorative (for example, silk-screened) part.

Temper

Harder tempers should be considered first since they often permit the use of thinner, stronger, and lighter material. Stiffer materials, however, require larger bend radii.

Recommended Slot Widths

All of the below slot widths are considered standard:

Screw size	Slot width
#2	0.093
#4	0.125
#6	0.156
#8	0.171
#10	0.218
¼	0.281

Folding

Folds are used on materials which can be bent back on themselves as shown in Fig. 4.16. To prevent cracking, especially on the flattened type, the folds are normally placed across the grain of the material (see section "Grain Direction"). The grain direction is, thus, usually specified.

Curling

Curling may be used for edges of sheet metal parts where rigidity requirements are greater than can be met by ordinary folding. The sizes of these edge curls should be in accordance with Fig. 4.17.

Crimping

The edges of large sheets may be crimped for extra rigidity and to prevent "oil canning." A typical crimp is shown in Fig. 4.18.

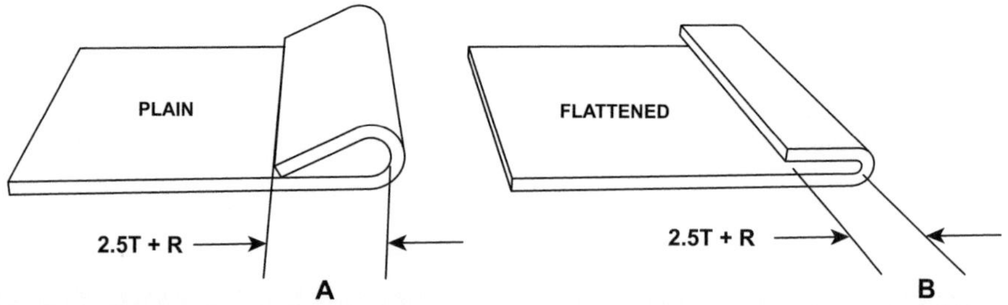

Fig. 4.16 Folding

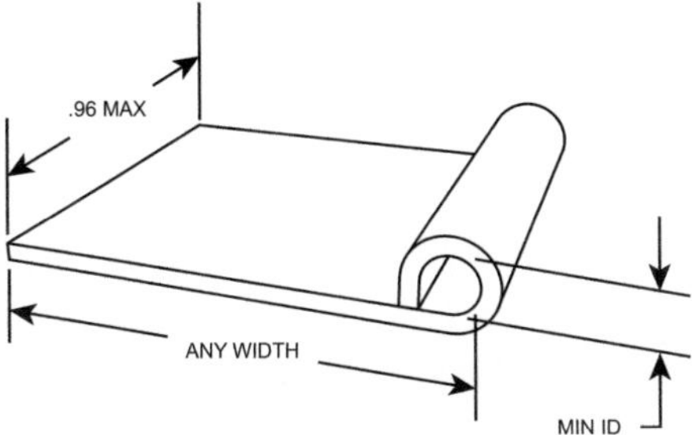

MINIMUM CURL DIAMETER	
THICKNESS	MIN ID
.001 thru .032	.12
.033 thru .040	.19
.041 thru .063	.25

Fig. 4.17 Curling

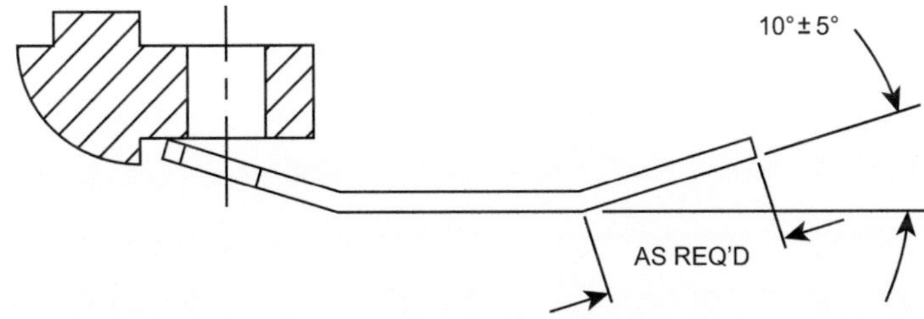

Fig. 4.18 Crimping

Joggles

Joggles are used to provide a step in sheet metal parts for lapping joints, etc. The width of the web or joggle allowance, dimension "L," should be a minimum of 3 times dimension "D," the depth of the offset. For brittle materials such as heat treated aluminum alloys, a minimum joggle allowance of 6 times the offset depth is recommended. Where the design requires that the offset depth exceed the material thickness, the 45° joggle should be used. See Fig. 4.19.

Dimpling

The most common application for dimpling is for countersunk head screws or rivets, see Fig. 4.20. Soft materials can be dimpled easily. Aluminum alloys such as 2024, 7075, and the harder tempers of stainless steels may edge-crack unless they are "hot-dimpled." Dimensional data for dimpling is given.

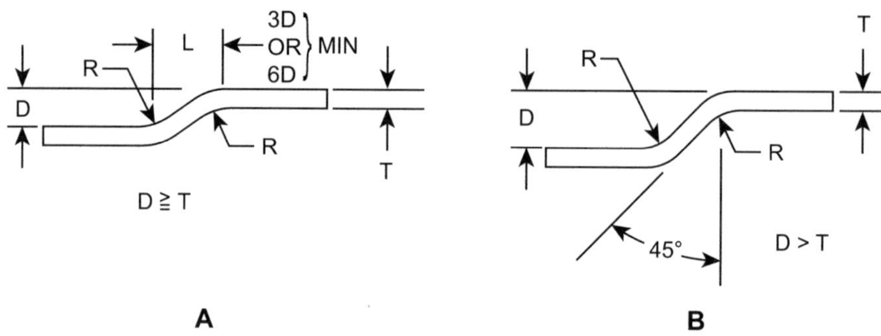

Fig. 4.19 Joggles

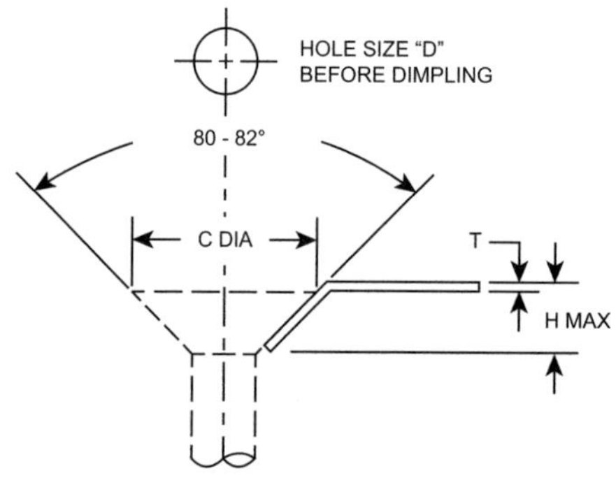

TABLE 1 - DIMPLING DIMENSIONS													
HOLE SIZE (D) BEFORE DIMPLING												C DIA +.010 -.000	H MAX
THK (T)	.016	.020	.025	.032	.040	.050	.063	.080	.090	.100	.125		
2 - 56	.047	.055	.060	.063	.067								
4 - 40		.055	.060	.063	.067	.070	.073						
6 - 32			.070	.076	.078	.082	.086	.094					
8 - 32				.089	.096	.102	.110	.113	.125				
10 - 32					.096	.102	.110	.113	.125	.140		.392	.195
1/4 - 20						.125	.144	.156	.166	.173	.185	.516	.250

Fig. 4.20 Dimpling

Countersinking

Where thicker material precludes punching or dimpling, drilling and countersinking may be employed to accommodate flat heat screws. See Fig. 4.21. This figure illustrates the dimensions involved with countersinking for an 82° included angle flat head screw (100° is also a commonly used flat head type). Note the "sample

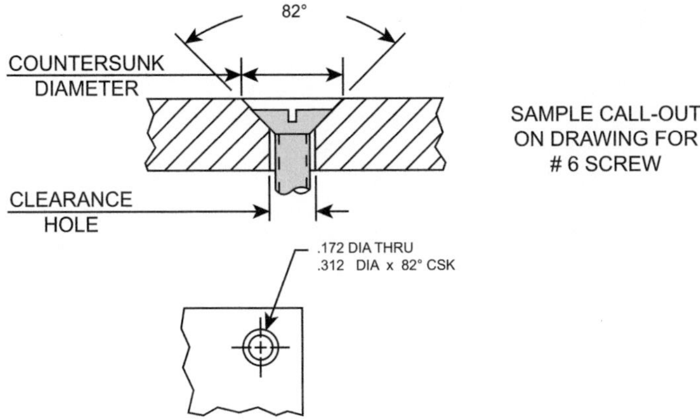

SIZE OF SCREW	MAX HEAD DIA	MAX SHANK DIA	NORMAL		PRECISION		ALL
			COUNTERSINK DIAMETER TOLERANCE ± .010	LEAST MATERIAL CONDITION	COUNTERSINK DIAMETER TOLERANCE ± .005	LEAST MATERIAL CONDITION	CLEARANCE HOLE ± .005
#2	.172	.086	.202	.085	.192	.075	.106
#4	.225	.112	.270	.102	.260	.092	.141
#6	.279	.138	.322	.118	.312	.108	.172
#8	.332	.164	.375	.147	.365	.137	.188
#10	.385	.190	.426	.162	.416	.152	.219
1/4	.507	.250	.545	.188	.535	.178	.281

Fig. 4.21 Countersinking

call-out" given for a #6 flat head screw ("precision" design). The call-out gives the (clearance) *hole diameter* and *countersunk diameter* (not the actual shank diameter and screw head diameter). Also note that the thickness of the material is normally at least the depth of the screw head.

Distortion from Bending

Figure 4.22 illustrates the distortion condition that occurs with forming operation. It is a particularly noticeable distortion when heavy material is bent with a sharper inside bend radius. It is "hardly noticeable" on material thicknesses less than 1/16 inch or when the inside forming radius is large in comparison to the material thickness.

The material on the inside of the bend is under compression, which results in this bulge condition on the edges. In addition, the edges on the outside of the bend are under tension and tend to pull in. This bulge or distorted condition is usually of no

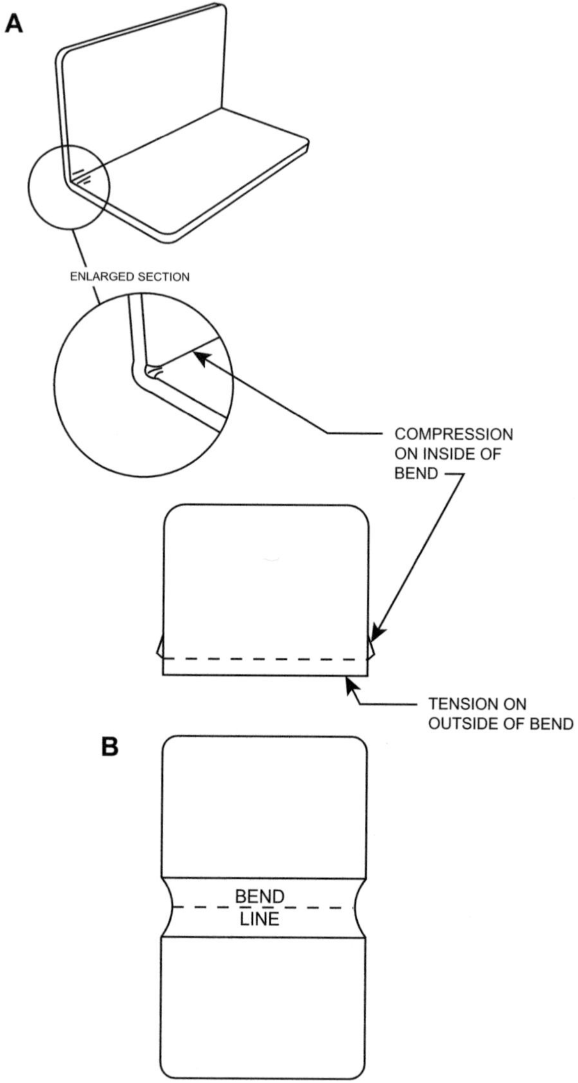

Fig. 4.22 Distortion from bending

concern and is accepted as standard practice. But, if this bulging will cause any interference with a mating part, then this should be referred to on the part drawing so a secondary operation can be considered to remove this interference. This extra operation may not require tooling but it will add to the cost of production.

Also shown in the figure is a blank developed to prevent interference resulting from bulge (without the extra production cost).

References

1. Equivalent rigidity or stiffness of material. Borg-Warner Chemicals, Inc Design Tips (1980). (Borg-Warner Plastics Division was sold to General Electric, and was subsequently sold to Sabic)
2. Ryerson data book. Joseph T. Ryerson & Son, Inc.
3. Ryerson products in stock & processing services. Joseph T. Ryerson & Son, Inc.
4. *Tool and Manufacturing Engineers Handbook*, Volume 3 Materials, Finishing and Coating, SME, Editors C. Wick, and R. Veilleux
5. *Guidelines for the Mechanical Design of Heat Sinks*, D. Burns, Thermalloy Inc.
6. *Plastic Part Design for Economical Injection Molding*, G. L. Beall, Prepared for Borg-Warner Chemicals
7. *Injection Molding - Theory and Practice*, I. Rubin, SPE
8. *Cycolac ABS product design manual*, Borg-Warner Corporation
9. *Industrial Painting*, Principles and Practices, N. Roobol.
10. *Casting Method Comparison Wall Chart*, Special Metals Supply Inc.
11. *Product Design for Die Casting*, Diecasting Development Council
12. *Designing in Zinc*, International Lead Zinc Research Organization, Inc.
13. *NADCA Product Specification Standards for Die Castings*, North American Die Casting Association
14. *Machinery's Handbook*. Industrial Press
15. *Dimensioning and Tolerancing*, ASME Y14.5 (or ANSI Y14.5)
16. *Geometric Tolerancing*, A Text/Workbook, R. Marrelli & P. McCuistion
17. *Geometric Dimensioning and Tolerancing*, Basic Fundamentals, D. Madsen

Chapter 5
User Interface

This chapter will add to our toolbox for electronic enclosure design. We'll start with a description of the input and output devices available to the designer to help the user to process data. A section on sensors that the user will use to "sense" this data will follow. We'll explore the five natural senses which will also include some "uncommon" senses. We'll add to the mix a discussion on the ever-present power sources that our designs must include. The discussion on how users will view and use our electronic enclosure will end with sections on the value of ergonomics, industrial design, and color, which will make our product both more useful and more enjoyable.

5.1   Input/Output

Many electronic products have methods to receive digital information, and a way to pass on digital information to another device. These products have methods the user can use to "input information" and a method that the user can "output information." This is not true for all electronic products. Some electronic products could be considered "active" (revising or processing information) or "passive" (little or no user interaction), and we'll explore some in more detail.

Examples of "active" electronic products are:

- Desktop or tablet computer – receives information in a variety of ways and outputs the same or different information in a variety of ways
- Cell phone – also receives information in a variety of ways and outputs the same or different information in a variety of ways

Many of the above "active" electronic products utilize either onboard software (or firmware) or utilize software from external sources to manipulate the digital information received and sent.

© Springer International Publishing AG, part of Springer Nature 2019

T. Serksnis, *Designing Electronic Product Enclosures*,
https://doi.org/10.1007/978-3-319-69395-8_5

Examples of "passive" electronic products are:

- Power supplies – can convert chemical energy into electrical energy or can convert alternating current (AC) into direct current (DC). Some of these power supplies can be very "smart" with printed circuit board assemblies, various sensors, and LED information.
- Cables – a pass thru or transformation of digital or RF signals.

We won't examine in depth the "passive" electronic products, but they do possess many of the attributes found in the sections on ergonomics and industrial design.

In an "active" product, there certainly must be some action taken by the user of the product, to "activate" that product. In its most rudimentary form, there would be an "on switch" and an "off switch." Of course, both the functions do not have to be activated by a switch that the user touches (physically); the function could be activated by voice or electromagnetic radiation within the electromagnetic spectrum. More on this "activation" will be explored in the following sections.

A designer of electronic product enclosures will be required to integrate into the product the sensors that will receive and transmit the digital information. They will be working with the rest of the design team to optimize sensor placement and usage within the product.

It's also a requirement for (active) electronic products to have a power source, and this is further explored in Sect. 5.3.

Be aware that technology moves very quickly in the area of user interfaces. Input devices like the computer mouse, only first "invented" (see Wikipedia reference on "Computer Mouse") in the 1940s and first placed in the Xerox Alto in 1973, has gone thru several innovate phases. This text will not be detailing the information on many of the standard user interface technologies.

Let's explore a (smart) cell phone to further define some of the input and output user interfaces that need design. A smart cell phone, for example, the Samsung Galaxy Core Prime, circa 2015, has several input and output interfaces:

- Earpiece – Allows listener to listen to a call
- Volume key – Allows adjustment of volume of the device sounds and audio
- Proximity sensors – Detects the presence of objects near the phone
- Front camera – 2 megapixel, takes self-portraits and records videos
- Power/lock key – Allows various modes of turning the device on or off
- Home key – Returns user to the "home" screen on the display
- "Recent" key – Allows user to display recent apps
- "Back" key – Allows user to return to the previous screen on the display
- Color touch screen display – 4.5 inch, WVGA, capacitive, 16 M colors, 480 × 800 pixel resolution. Provides means for QWERTY keyboard input.
- LED flash – Illuminates subjects in low-light environments when taking a photo or recording video
- Headset jack – Connects an optional headset
- Microphone – Records audio and detects voice commands
- Rear camera – 5 megapixel, autofocus, f/2.6, takes pictures and records videos

- Speaker – Plays music (MP3) and other sounds
- USB charger/accessory port – Connects the charger/USB cable and other optional accessories
- SIM card – Provides user subscription details
- MicroSD or MicroSDHC memory card port – Allows installation of optional memory card to expand available memory space
- Battery compartment – Allows installation and replacement of 2000 mAh Li-ion battery
- Accelerometer sensor
- Temperature sensor
- RF antennae and receivers – Wi-Fi 802.11/Bluetooth/GPS/Glonass/BeiDou/ NFC
- Messaging – SMS/MMS/Email/IM
- Vibrator – Sound and vibratory message alert

As shown by the above listing of input/output interfaces, this is indeed quite an arsenal of available technology for the designer to understand so that they can best locate them in the product for maximum utilization. An interesting discussion on how this many tecnologies ended-up in one place (in this case, the iPhone, see Ref. [5]. Ref. 5 The One Device, the Secret History of the iPhone, B. Merchant, Little Brown and Company, 2017.

5.2 Sensors

The sensors listed in the cell phone in Sect. 5.1 generally stimulate (or activate) the three human senses of:

5.2.1 Sight (Vision)

The eye is sensitive to visible light waves in the frequency spectrum of between 400 and 700 nanometers. Our sight provides us with a huge bevy of input and output devices that are in common usage. Displays use digital signals to energize individual pixels that are either "on" or "off" and have black, white, and entire palettes of color. Displays let us "see" or "convey" information of all sort. The displays can change their pictures at high rates of speed that conveys motion, indeed, either at actual (real), slowed, or increased speed rates.

Besides displays, other light sources are used to convey information. An LED can be used to signify "on," "neutral," or "off" or indeed be grouped together in circular patterns to form an "information-rich" object that is called a stoplight.

Labeling (letters, symbols, graphics) also provides information to our sight and thus also is a part of our gathering of "visual information."

Sensors have also been created that are outside of the visual frequency spectrum. Longer wavelength (infrared) and shorter wavelength (ultraviolet) sensors extend the range of our ability to sense visible light.

5.2.2   *Touch (Feel)*

Touch is used quite extensively to convey information. Physically pressing a key (such as on a keyboard) sends a digital signal that is received and processed on. Keyboard mice use touch to pick, scroll, or move cursors that again convey information. Physically touching a touch screen chooses information that is received and processed on.

Keys, pads, and switches all can provide a way to either convey or receive information.

Another way that touch (feel) is used is when vibratory (or shock) motion is produced that announces some information to us.

5.2.3   *Hearing (Auditory)*

We use our hearing quite regularly to get information. Technically, our hearing is actually a "touch mechanism" in that it is sound (pressure) waves that touch our hearing mechanism. This hearing mechanism contains the three smallest bones in the body, the malleus, incus, and stapes (the hammer, anvil, and stirrup). Hearing and touch are types of mechanosensations where molecules move and are transduced into nerve impulses that are perceived by the brain. Frequencies capable of being heard by humans are between 20 Hz and 20,000 Hz.

An interesting aspect of the sense of hearing is the field of speech recognition (SR). According to Wikipedia, "SR is the inter-disciplinary sub-field of computational linguistics which incorporates knowledge and research in linguistics, computer science, and electrical engineering fields to develop methodologies and technologies that enables the recognition and translation of spoken language into text by computers and computerized devices such as those categorized as smart technologies and robotics. It is also known as "automatic speech recognition" (ASR), "computer speech recognition," or just "speech to text" (STT)."

It is clear that the use of SR is indeed a powerful tool. It is being used as an information provider (to the user) and being used by the user to provide information to others.

5.2.4   *Antennae and Receivers*

Transmitters transmit electromagnetic waves that have both frequency and amplitude. Antennae receive those signals and process them into information. The FCC and worldwide frequency allocation regulatory agencies regulate specific frequencies so that they do not overlap (interfere) with each other. Common frequencies used for communication range from 300 MHz to 3 GHz. They are used in:

- Mobile phones
- Wireless LAN
- Bluetooth
- GMRS radios
- GPS

A rather interesting antenna/receiver "pair" is the recent use of brainwaves (whose speed varies from infra-low (<0.5 Hz) to gamma waves (42 Hz) to accomplish all sorts of things. Reference [3] chronicles the work of Rajesh Rao. A brain signal was sent over the Internet that moved the hand of another person at that location. This realm of Human Machine Interfaces (HMI) has several startling examples:

- Composition and playing of music – creating music with thoughts
- Screen mobile phone calls – filters incoming calls by monitoring the state of the user's brain
- Create a 3D object – objects created by thought alone (Thinker Thing)
- Driving a wheelchair and a car – thought-controlled vehicles
- "Bionic" limbs – prosthetics interface with the wearer's neural system

5.2.5 More Senses

Two more human senses are not used in the cell phone (above) but are actually now utilized in some software and hardware applications for an iPhone (see [1, 2]).

- Smell (olfactory)
- Taste

Four more senses that some people believe that humans possess are:

- Thermoception – sense of temperature
- Nociception- sense of pain
- Equilibrioception – sense of balance
- Proprioception – sense of body position

The last two senses listed above are utilized in a product that has lost some "traction" recently, the Segway. However, the traction was regained with the introduction of the self-balancing scooter, the hoverboard. Gyroscope sensors are used in these products to take advantage of a person's sense of balance and their sense of body position, to help provide transportation for an individual.

Thus, the five natural senses, which were known to serve as a way into the brain, and with language and gesture as the channels out, our user interfaces have been augmented.

5.3 Power Supplies

Power supplies are being listed as a separate section in this chapter on user interfaces, as they are broadly speaking, an input.

All electrical products need a power source. A *mechanically* powered flashlight is a flashlight that is powered by electricity generated by the *muscle power* of the user, so it does not need replacement of batteries or recharging from an electrical source.

Sometimes the power source is "external" (or separate) from the product, as is the case of the flashlight above or when the product is powered by standard household alternating current (AC). Another "external" power source is solar power, which is converting the energy from the sun into direct current (DC). In most other cases, the power source is "internal" and can be:

- Rechargeable battery
- Unrechargeable battery
- AC to DC converter

If the product is to be powered by a battery, there will normally be a way to insert the battery (door and contact terminals). That is, user access is provided to the battery. Battery (chemical) technology advances with the times in an ever-increasing attempt to optimize the "energy density" ratio of more power for less weight. The current solution of using Li-ion chemistry is pervasive in automobiles, power tools, and computers.

Power sources have special considerations in the areas of cooling (see Chap. 8), EMC (Chap. 9), and safety (see Chap. 10).

5.4   Ergonomics

The terms "ergonomics" will be defined here, with the next Sect. 5.5 used to delineate "industrial design," although certainly they are closely related.

I'm going to copy (quote from) what I feel is the best definition and discussion of these terms here, and this is from [3]:

> It will be noted that the term "ergonomics" can be used as a subtitle for this handbook because "human factors engineering" and "ergonomics" have been used interchangeably in discussing the general topic of product design as it relates to user efficiency. In spite of some observers' attempt to define the two terms differently, in actual fact those who work under the guise of "human factors engineers" and those who call themselves "ergonomists" actually use the same basic information and perform the same kinds of work with respect to design. Their expertise and purposes are for all intents and purposes identical. The only difference one might perceive is that in the United States the term "human factors engineering" is used more widely, while in other countries the term "ergonomics" has been more predominant. It has often been noted, however, that the term "human factors engineering" is closely associated with U.S. military design programs, whereas in other countries "ergonomics" has been closely allied with industrial human factors work.

I'm certainly not going to get into many specifics on this well-discussed subject of ergonomics but will highlight some aspects as it relates to the design of consumer products and electrical products in general.

In the area of mechanical design of electronic product enclosures, there is much data available in the areas of:

5.4.1   Visual Displays

General guidelines include:

- Use the simplest display concept – reduce time to read and interpret
- Use the least precise display format – response time goes up as user has to be precise

- Use the most natural or expected display format – reduces response time
- Use the most effective display technique – match the display to the operator
- Optimize for visibility/conspicuousness/legibility/interpretability

5.4.2   Controls

Important considerations in control selection, design, and use:

- Use control that is operable in terms of the natural motions of the arm, wrist, finger, foot, etc.
- Provide feedback so that the operator knows at all times what their input is accomplishing.
- Provide a good balance of control resistance so that spurious input is dampened, but operator is not quickly fatigued.
- Position the control to minimize excursion.
- Choose a size and shape of the control interface to maximize user control and minimize inadvertent actions.
- Choose a surface texture for the control that allows smooth transitions or firmness of grip.
- Consider the choices between one-hand vs. two-hand operation.

5.4.3   Fasteners

Guidelines for the design and selection of fasteners are:

Use the minimum number of fasteners commensurate with the requirements of securing components. Use tongue and slot (or similar techniques) to reduce the number of fasteners.

- Use "captive" fasteners to avoid loss of all or part of the fastening device.
- Work space should be provided around a fastener for fingers and tools.
- Standardized fasteners should be used wherever possible to minimize stocking requirements and the number of different tools required.
- Use fasteners that are operable by commonly used rather that special tools. An exception to this would be utilization of a "tamper-proof" fastener in areas where the user might experience a safety or servicing issue.
- Use the same type and size of fastener consistently for a given application. Care should be taken if fasteners with different threads are used in close proximity to each other so as to minimize the chance of incorrect fastener being used. First choice would be to use NO fasteners on a product (certainly a minimum number). Second choice would be to only use a single fastener size and length.
- Consider how a worn or stripped fastener can be removed before selecting it – use screws rather than rivets.
- Where vibration could result in loosening of a fastener, provide locking mechanism (lock washer, thread-adhesive, or even a safety wire).
- Care must be taken when springs are being installed to avoid safety issues in use or installation.

5.4.4   Physical Response Measurement

Much ergonomic data exists [3] in the areas of:

- Visual response
- Auditory response
- Tactile response
- Acceleration response
- Vibration response

5.5   Industrial Design

Industrial design (ID) is the professional service of creating and developing concepts and specifications that optimize the function, value, and appearance of products and systems for the mutual benefit of both user and manufacturer. This is the definition from the *Industrial Designers Society of America.*

One cannot discuss the subject of industrial design without mentioning the term "attractiveness." Again, I'd like to quote [3] on the subject of "attractiveness":

> Although most manufacturers and designers are fully aware of the importance of product attractiveness to *salability*, and thus pay more attention to this aspect of product design than to any other, attractiveness is an *elusive* variable when one tries to define it. A critical comment is in order also: Although attractiveness is an admirable and desirable trait in product conceptualization, it is extremely important that, in an effort to make something attractive, the designer be careful not to compromise *product safety*, *operability*, or *ease of maintenance*. (Italics, mine)

Reference [3] goes on to state:

> Since attractiveness is truly a human factors consideration, it is important to recognize that it may be a transient phenomenon; i.e., what one considers attractive this year is not necessarily considered attractive several years from now. This point is made to refute the argument that the aesthetic function is the most important consideration in consumer product conceptualization; i.e., a product that works well can continue to do so over many years and through many different packaging treatments, but only if the initial conceptual objective was to *make the product work well*. It should also be emphasized that a product can be made attractive after the objectives of *operability* and *maintainability* have been developed. (again, italics, mine)

One of the best examples of product industrial design that is well done are those products released by the Apple Corporation. Apple has *defined* what products should look like in the area of MP3 player, laptop computer, and certainly the smart cell phone. "Well done" can be said easily by *any* definition of sales, iconization, and overall effect on the products that followed from competitors.

It varies from product to product and from company to company as to how large an effort is made to enhance the industrial design aspects of a product. Apple would be an example of the highest regard that a company can attend to this factor as it reflected in their products and the size of the effort that they place upon ID. Smaller efforts abound in the industry as some products are designed with a one-person combination of mechanical designer and industrial designer.

Many companies create a "design standards manual" (or set of guidelines) that attempts to:

- Unify the appearance of selected areas of the products either across product lines or across *all* products from that company.
- Convey to customers, employees, and the business community the ability to recognize those company products thru unique and consistent product features.
- Convey thru product appearance the sense of quality and innovation that lies within.

5.6   Color

One of the most "visual" examples of an "industrial design tool" is the use of *color*. Color science is an extremely complex field. It is possible to describe colors in many different ways; some of which depend on how the color is created (e.g., from a light source directly), reflected (as from a printed page), or shown on a computer display. Some descriptions are objective measures of light itself (e.g., the peak wavelength), but others are keyed to knowledge about how a typical human eye perceives colors.

Three primary measures of color are (from [4]):

- Dominant wavelength – a single frequency
- CIE xy color coordinates – a pair of numbers
- RGB color components for computer monitors – a triad

(Printing of colors as with CMYK systems is not addressed.)

Another set of colors are the 216 standard "web colors" often used on web browsers. These are derived by combining each of six different intensities each of red, green, and blue in every possible combination. In HTML, colors can be expressed as RRGGBB hexadecimal numbers (each digit in the rage 0 thru 9 or A thru F) to represent a range of 0 to 255 for each color (in some cases listing two web colors when the range limit falls between two colors).

References

1. *Harvard Scientists Send the First Transatlantic Smell via iPhone, Alyssa Bereznak, Yahoo Tech, 6-17-2014.*

Harvard scientists successfully transferred the first scent from Paris to New York on Tuesday morning via an iPhone app. The champagne and passion fruit macaroon-scented message was transferred via a new communication platform called the oPhone.

It works like this: A custom-made app allows you to take a photo of something and "tag" it with a few aroma notes (from more than 3000 scents). These smells – which range in category from "Paris Afternoon" to "Plantation" – are transferred via a pipelike smelling station called an oPhone Duo and are controlled by an iPhone app called oSnap.

When you send an oNote, your recipient will click a link that leads him to a photo, as well as the specific aromatic notes you have chosen. When connected to the oPhone Duo, the hardware piece, it'll emit slight scents from two separate pipes to be smelled in conjunction with the message. Otherwise, the app will just offer some vivid description of the scent your sender is trying to convey.

You don't have to own the oPhone hardware, which starts at $149 on the company's Indiegogo page, to send or receive a smell. Anyone without the contraption will still be able to tag images using the oSnap app (out in the App Store now) and mark it with around 16 different high and low notes. Currently, user creations range from "Lady Gaga" to "My Room" to "Smoky Beach."

2. *Introducing OpenTable Taste: 4D Lickable Technology Allows Diner to Taste Dishes in App, Olivia Terenzio, 4-1-2016.*

At OpenTable, we are always looking for new and innovative ways to connect people with amazing dining experiences around the world. We're especially excited to help diners discover new restaurants and help them experience restaurants before they ever step in the door.

For years OpenTable diners have requested a way to get a taste of a restaurant's food before booking a table. Our research shows diners have a biological reaction when they see evocative food photography – food porn leads to higher levels of endorphins and oxytocin and, ultimately, pleasurable experiences.

Today, OpenTable is thrilled to announce the release of OpenTable Taste, a groundbreaking new way for diners to experience the food photos they love. With the OpenTable iOS or Android app, you can now sample a restaurant's food directly from their phones – all you have to do is lick. Our engineers have developed an advanced algorithm to map out tongue taste buds, then engage with them to recreate the flavor profiles in your brain. This technology sends signals to your brain, meaning for the first time in history you can taste food photos through your phone.

How It Works

Diners can experience OpenTable Taste technology in three easy steps. It's easy (and tasty) as pie!

First, install the latest version of the OpenTable iOS or Android free app; U.K. only download for iOS or Android here.

Next, find a scrumptious photo and give it a little lick.

Finally, book at that restaurant or pick another to try. (*Note: Multiple licks may be necessary.)

3. Woodson WE (1981) Human factors design handbook, information and guidelines for the design of systems, facilities, equipment, and products for hunan use. McGraw-Hill Co., New York.

4. Optoelectronics application manual (1977) Hewlett-Packard Company.

Chapter 6
Assembly and Service

We want to make the product as easy to assemble as possible. But, unless we think about that at the earliest stages of design, it will not happen – this takes planning! This chapter will span the region of product design that begins with the ability of a product to be "manufacturable" and ends with the ability of a product to be "serviceable." This is a bit incongruous in that "design" cannot (should not) be separated from its ability to be both assembled and serviced. The chapter will start by defining the various assemblies that the designer will be concerned with and the key assembly concerns. We'll provide some assembly guidelines. The issues of the needs for assembly tooling and assembly testing will be discussed. We'll end with a section on service considerations. An appendix is included that gives a bit more detail on how various hole sizes are determined to help in the assembly process.

6.1 Design for Assembly

First of all, some terminology will be used in this chapter. This chapter is concerning *the assembly* of multiple individual parts into a sellable product. The manufacturing (or fabrication) of individual parts was covered in Chap. 4. Design for assembly (DFA) has been identified by corporations as a critical and measurable aspect of a product. Boothroyd Dewhurst's design for assembly software has been used in industry to reduce part count and assembly time for over 25 years. This software puts a metric on each step in the assembly process and helps identify ways to reduce the total time and cost. Reference [1] shows a case study of this particular DFA program for Hypertherm, Inc. Reference [2] is a paper written by B. Huthwaite on the subject.

© Springer International Publishing AG, part of Springer Nature 2019
T. Serksnis, *Designing Electronic Product Enclosures*,
https://doi.org/10.1007/978-3-319-69395-8_6

I'm going to make a distinction between:

- Final assembly – where the product is assembled before customer shipment (or, customer warehouse). Final assembly includes testing required to insure compliance to specifications for the product.
- Subassembly – any subassembly done external to the place where final assembly takes place. There may *also* be subassemblies done where the final assembly takes place, and these assemblies may be done apart from the final assembly flow of parts. Thus, subassemblies can be produced either externally or internally.

The assembly of individual parts entails several entire departments of a corporation as we are generally concerned with all of the separate components of the supply chain network, including:

- Parts documentation, commodity identification, build demand
- Parts quoting, ordering, quality control, and shipping to stock (or directly to assembly)
- Approved vendor identification
- The entire organization responsible above can be termed "manufacturing" and is also known as "operations." The individuals working together on the assembly process are:

Design engineers
Manufacturing engineers
Service engineers
Test engineers
Project management
Purchasing managers
Document and change control

Just as an individual part designer must know, "Cost" (Chap. 4), the assembly of those parts, must be thought of as "Cost (Chap. 6)" which I will further explain. For Cost (Chap. 6), all of the trade-offs between cost, time, and conformance to specification *for assembly* must be known. Just as we needed to know the cost of the individual part, we need to know the cost of assembling that part.

The first item to identify for the *final* assembly process is to determine whether some of the parts can be subassembled, before they are ready to be in the "final assembly" process. It's very possible that some subassemblies can be assembled before they reach the final assembly stages (either by a different assembly supplier or your in-house final assembly operation).

Key items to identify concerning assembly are:

1. How many assemblies are to be produced over the lifetime of the product? And how many assemblies are to be produced to the expected sales forecast for a specific period? We would need to know the "reasonable" production schedule. Assembly lines look very different as quantities go from one to thousands per month.

2. Generally, how long will it take to build one assembly? Again, the scope of work changes dramatically if the product is something as large/complicated as a commercial jet aircraft or as "finite" as a cell phone.

3. How "automated" is the assembly process expected to be? Are some operations to be performed by robotics? We'll discuss more about tooling in Sect. 6.3, but we need to have a sense of the assembly/fixture tooling budget to be able to fully plan the assembly process.

4. How much testing will be needed during assembly stages, and where are the reasonable stages to test the product? Is 100% testing needed to obtain the highest reliability? Again, if the product is a heart monitor, that may have different reliability requirements than a cell phone.

5. Where will individual parts (or, purchased subassemblies) flow onto the assembly area? When will these parts arrive for the assembly process?

6. How many individuals are needed for an assembly period? This certainly is linked to knowing how long it will take to build an assembly.

7. How large of an assembly area will be needed to assemble/test the product?

8. Should the final assembly be done in-house or contracted out?

The designer's thought process must include:

- Design of the individual part
- Design of the completed assembly
- How each individual part is assembled *into that completed assembly*

The above can be done most successfully, if the above is done "all at the same time," that is, the *assembly process* is kept in mind as parts are being designed. Exceptions can occur when a designer doesn't "burden" themselves with any assembly technique (at first) just to "blue-sky" a design. However, eventually the designer must "double-back" to that design's assembly process.

6.2 Assembly Guidelines

The following is a list of some items that will help make a productive assembly process:

1. Attempt to assemble "from a single direction." That is, the assembly should "build" from one direction, the top direction (Z-direction). The top direction takes advantage of gravity. The more times the assembly "flips," the more time it will take to build. Assemblers (people) can see and inspect what they have done more easily from the top direction.

2. Reduce the number of parts in the assembly. If two parts can be combined into a single part, attempt to accomplish that. Be careful of Cost (Chap. 6) in this respect. Can four separate brackets be combined into one "super-bracket"?

3. Use a minimum of fasteners in the assembly. If you use a fastener in the assembly, attempt to use the *same* fastener (same thread size and same length). Use

the same head drive (for applying torque/tightening) for the fasteners. It will be beneficial to *not* have similar lengths of fasteners, such as both a #8-32 × 0.31 long and a #8-32 × 0.38 long, as this will lead to assembly errors. Try to standardize on fasteners, not only for this assembly but for other assemblies. Standardize on *either* metric or American national sizes.

4. Reduce the "stack-up" or number of separate items needed for a fastener position. That is, instead of a screw + flat washer + lock washer, look to combine those features into a screw with an integral large bearing face and a self-locking feature (such as a "SEMS" or lock patch), "Cost" (Chap. 6).

5. Avoid glues or adhesives in assemblies unless the adhesive application has tightly controlled procedures. Repeatability (and quality control) of the assembly is certainly a question to be considered. Protect against the inhalation of fumes in these procedures and also protect for any other human safety issues relating to the application of adhesives. Material safety data sheets (MSDS) must be available to state any considerations with the usage of chemicals. Consider out-gassing of the glue or adhesive during the life of the product and how that out-gassing product will react with the other parts of the assembly. Glues or adhesives are *not* considered serviceable and certainly cause recyclability to be limited. Glues and adhesives must also be investigated for their long-term reliability in temperature extremes, humidity, UV, etc. Most glues or adhesives have environmental disposal issues and shelf-life issues.

6. Clearance holes need to be sized properly to allow tolerances between bolt patterns to line up (see Figs. 6.1 and 6.2 and Appendix 6). Several charts are available to specify the hole size needed for different "patterns" of holes, whether the fastener is in a "fixed" or "floating" condition and the screw nominal size. The hole size generally is as large as possible while still maintaining enough bearing surface for the head to apply enough force in the screw joint. The use of slots can help in an assembly if a pattern of holes will not line up due to increased tolerances in fabrication processes.

7. Internal cabling needs proper cable "management" (see Fig. 6.3). That is, the cables need to be tied-down in defined locations so that they are routed in a consistent manner. Installation of a cable should be such that it is well-marked for polarity of the mating and cannot be mismated. Connector mating should have a mechanism that insures that the mating will be locked in place after assembly, that is, it cannot become unmated in use (due to shock or vibration) but can be disassembled for service. Cable lengths need to be long enough to reach their mating when the assembly is "open"; however, they can't be so long as to allow themselves to be "pinched" once the assembly is closed. Look to have "standardized" cables (same connectors, same amount of pins, same lengths) in the same assembly as this can be cost-effective and avoid confusion.

8. Include in the individual part design some "assembly cues." This would include any marking on the parts which helps assembly, such as part marking that states *this end up*, or *connect* to P1. Color coding can also be used. Look to design parts that *cannot* be assembled incorrectly. Hole patterns can be made unsymmetrical (on purpose) so that the part cannot be assembled incorrectly. Also,

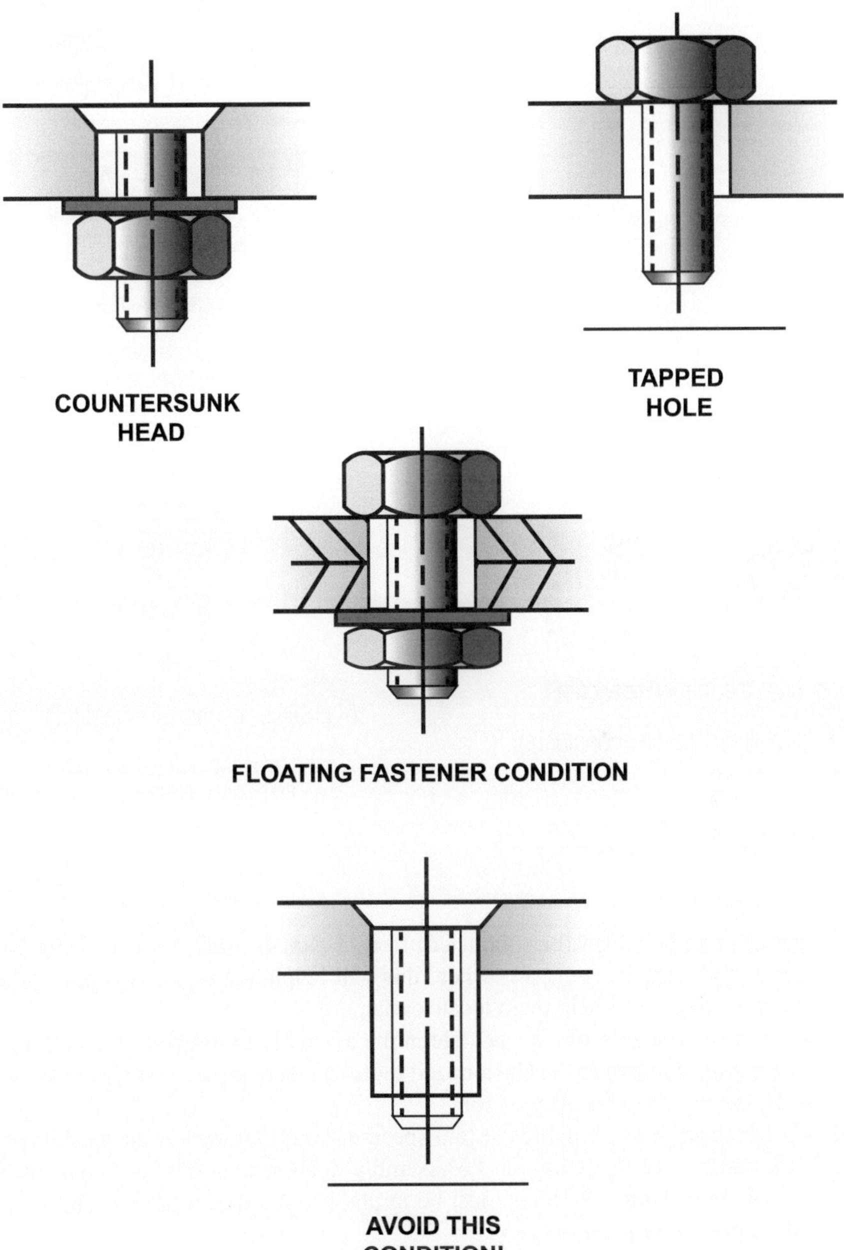

Fig. 6.1 Fixed vs. floating fastener condition

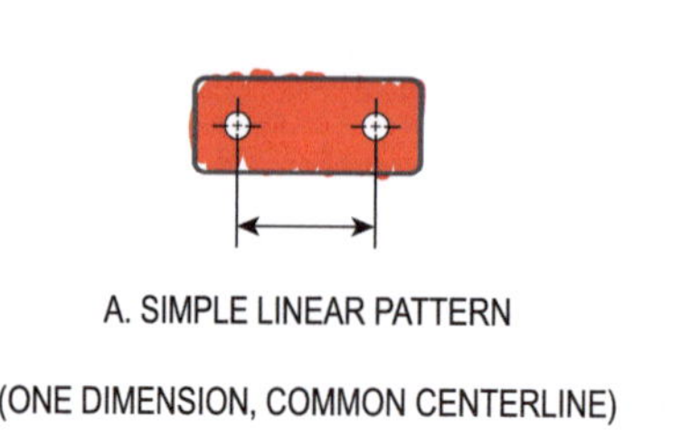

A. SIMPLE LINEAR PATTERN

(ONE DIMENSION, COMMON CENTERLINE)

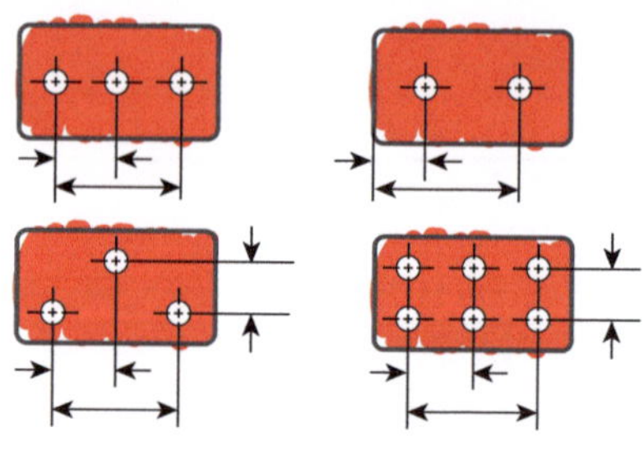

B. COMPLEX LINEAR PATTERN

(TWO OR MORE DIMENSIONS WITH COMMON
CENTERLINE OR ONE DIMENSION IN ONE
COORDINATE WITH TWO OR MORE DIMENSIONS
IN THE OTHER COORDINATE)

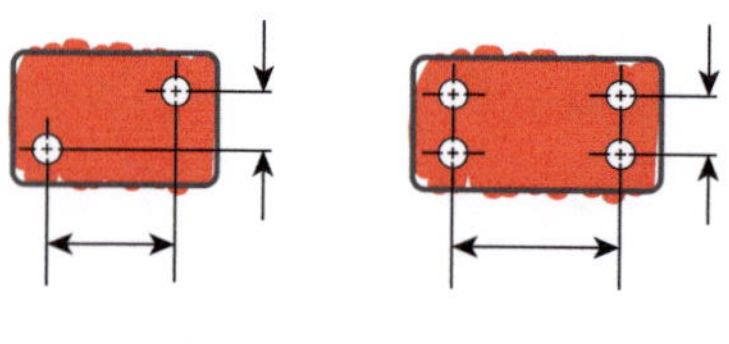

C. SIMPLE COORDINATE PATTERN

(ONE DIMENSION IN EACH COORDINATE)

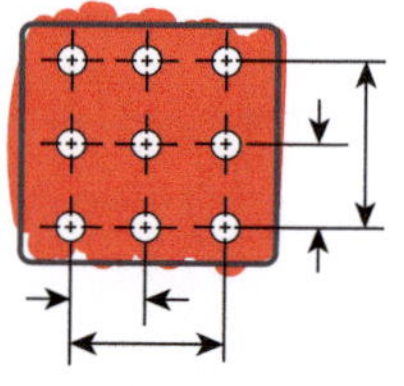

D. COMPLEX COORDINATE PATTERN

(TWO OR MORE DIMENSIONS IN
BOTH COORDINATES)

Fig. 6.2 Types of hole patterning

features can be added (or subtracted) from a part to insure proper assembly. Add a "key" (groove) to parts where this will help align or locate a part. Add chamfers to parts to help them locate easier.

9. Insure that enough room is available in the assembly to torque-tighten all fasteners with standard tools. Customized tools can be designed if this is not possible, but this must be planned for.

10. All fasteners must be tightened to a specified torque (range of acceptability). This either must be stated on the assembly documentation or be a part of a specified procedure. Systems must be in place to monitor torque settings and calibration of torque tools.

11. Protect the cosmetic surfaces of the assembly against damage that might occur during the assembly process. In fact, these surfaces need protection all the way to the customer.

12. Make sure all sharp corners and burrs are removed from individual parts so that assemblers are protected from injury.

13. Avoid assembly steps that specifically make the disassembly (servicing) "impossible." For example, if a shaft is put thru a hole in a plate, and then has a

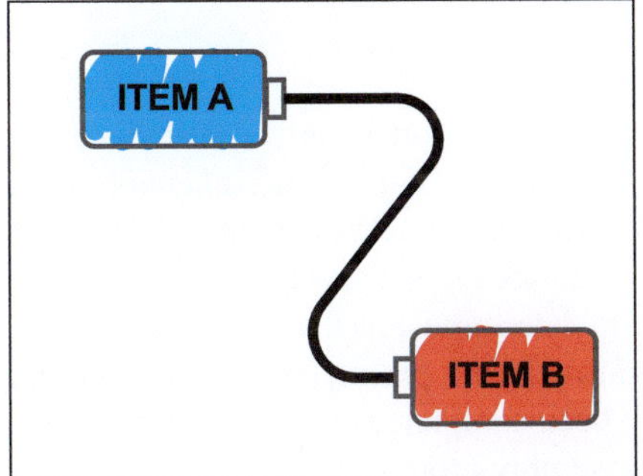
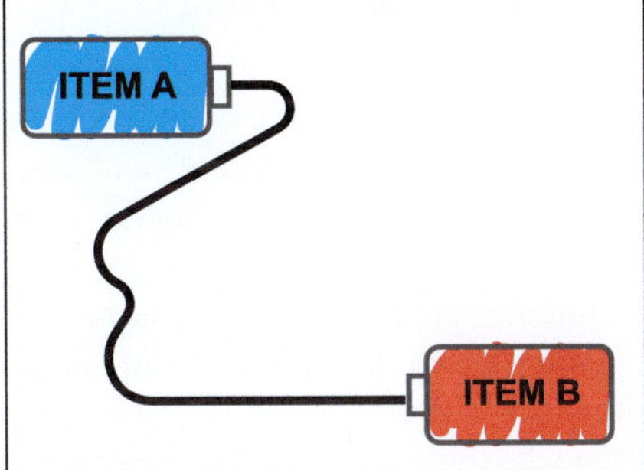

WITHOUT CABLE MANAGEMENT
(INCONSISTENT ROUTING)

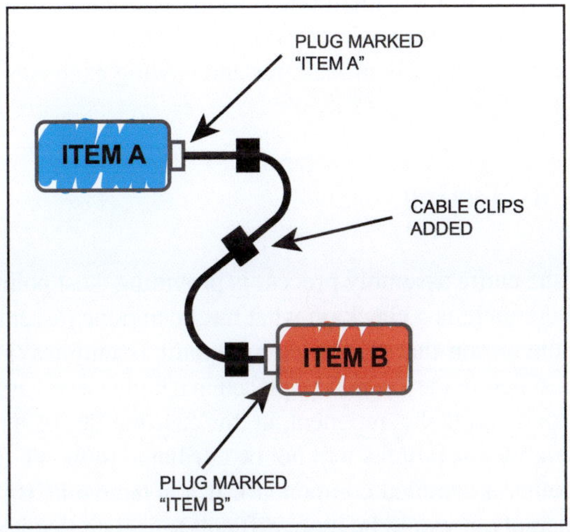

WITH MANAGEMENT

Fig. 6.3 Cable management

part welded (or riveted) to it, so that now the shaft cannot be removed thru the same hole, this becomes an "inseparable assembly" which proves troublesome in service or any other disassembly.

14. Look to add to the assembly process any fixturing which can make an assembly step both repeatable and within specification (see Sec. 6.4 on Tooling).

15. Provide lifting or counter-balancing apparatus to keep the assembly safe and ergonomic.

16. Look to add any automated data collection systems such as bar coding/ optical scanning that will provide work-in-process control and accountability.

6.3   Assembly Tooling

As with the fabrication of individual parts, tooling can greatly cost-reduce the assembly process. This tooling can save assembly time, but its largest benefit might be that it can "insure" that quality control is maintained. As with any cost/benefit analysis, the cost of the tooling needs to be compared to the savings gained with the addition of the tooling to the assembly line to determine a payback period.

Many times, the assembly line is brought up under prototype or preproduction timelines where the assembly tooling hasn't been completed (or, considered as yet). Thus, the more analysis that goes into the assembly process well before producing "many" assemblies, the sooner this tooling can developed (or, with enough planning, potentially the need can be eliminated).

Consideration of assembly tooling not only applies to the final assembly but also applies to any subassemblies, including those subassemblies that are done by outside fabricators. Usually, this is already considered by the outside fabricator, but not always. So, these subassembly procedures and costing exercises need scrutiny.

6.4   Assembly Testing

A large cost of the entire assembly process is providing "test points" in the assembly line process where there is a check on what has been done (assembled) to that point. This would be the reason that printed circuit board assemblies (PCBA) are checked for their operation before they would be placed in a higher-level assembly. The *sooner* it is found that an individual component, in this case the PCBA, is defective, then the subsequent assembly test failures will not be attributed to the PCBA (as it could be a failure from another assembled component). If a *defective* PCBA was placed in the higher-level assembly, it would be more difficult to isolate that it was the PCBA that was defective in a *subsequent* test point. It would then require disassembly (removal) from the higher-level assembly, and this would be highly unproductive.

So, every time that a component is added to an assembly step, that component needs some level of "testing" to insure that it is within specification *before* it is placed in the final assembly. An extreme example would be to test a cell phone only after it was *completely* assembled. If it did not power up, one could suspect a number of potential failures, possibly in just about every component within the cell phone. However, if only one per million (as an example) complete assemblies had a failure, it might be justified to only test the cell phone after it was completely assembled, as one could **scrap** the one per million assemblies as it doesn't pay to determine the actual cause of the failure. Warning: Customer Satisfaction is very important, so any failures can cause the overall quality of the product to be lowered. At some point (and maybe it would be *ten per million* in our example), it would be cost-justified to see if the cause of the failure could be determined.

Assembly testing can be as automated as it needs to be considering all of the previously determinations of cost. If a highly automated testing station can insure an acceptable level of assembly quality, the cost for that testing station can be justified.

Where and how the assembly will be tested is an essential part of the overall assembly process and must be planned for as such.

6.5   Service Considerations

How a product will be serviced and warranted is a very important component of the entire overall cost of the product. Usually some warranty "period" is either stated which is included with the cost of the product, or a warranty is a separately sold purchase of the customer that is valid for some period of time (See Ref [3]). Some service options available to consider:

1. No customer service is permitted. The customer calls for a service technician to either visit their site or the product is shipped back to the seller's service center (or, walk-in service center).
2. Some service is permitted by the customer. An isolated module of the product is shipped back to the seller's service center. A replacement module is shipped back to the customer for reassembly. Proper packing material and instructions are essential to make this an effective method.
3. The product will not be serviced by the customer or the seller. If defective, it (the entire product) will be scrapped by the customer. A replacement will be sent to the customer if the product is under warranty.

If a product is under warranty, and service is performed, it must be clearly stated to the customer as to what parts are new or used and what is the applicable warranty remaining on that original product.

Some products are designed that need scheduled service and/or maintenance, such as automobiles or copiers. I'm not going to address this type of servicing which is regularly scheduled. Most small electronic products would not require regularly scheduled servicing with the possible exception of battery removal and installation.

Whether the product is serviced by the customer, or seller, the product must be designed so that during the servicing:

1. It is safe and nonhazardous to do so.
2. There is a complete set of disassembly/assembly instructions to follow.
3. There must be a cost-effective determination of the cause of the defect or exactly what part needs repair/replacement.
4. All servicing must be done following the "approved" assembly and test instructions that were followed during the (original) build.

Most electronic products would be designed knowing that servicing is cost-prohibitive. Besides the very costly defect identification, you have a loss of customer satisfaction. In the age of social media, a single dissatisfied customer can reach millions of people rather quickly.

Another problem that occurs is identification of "customer abuse." Products can be warranted for particular environmental levels, but it is difficult to determine whether customers have exceeded these environmental levels. For example, a product may be "guaranteed" (tested) to still work after a one meter drop. But customers may claim that the product broke (even though they dropped it at a distance of 2 meters). More on this aspect will be explored in the next chapter on "Product Environments."

Appendix 6 Clearance Holes for Assembly

Definitions

Fixed fastener condition:	A condition where the fastener is free to float in a hole in one member and constrained from moving by a feature (such as a countersink or threaded hole) in the other member.
Floating fastener condition:	A condition where adjoining members each have holes that provide clearance between the hole and fastener.
True position tolerance	where *clearance equals* the minimum hole size minus the maximum screw size.

$Z1$ = true position tolerance diameter, Item 1
$Z2$ = true position tolerance diameter, Item 2
Zt = total true position tolerance diameter = $Z1 + Z2$
$H1$ = minimum hole diameter, Item 1
$H2$ = minimum hole diameter, Item 2
F = maximum screw diameter

Coordinate tolerancing has location tolerances determined by equations based on the fastener condition *and* the type of *hole pattern* (see Fig. 6.2).

$Z1$ = +/− position tolerance, Item 1
$Z2$ = +/− position tolerance, Item 2
$H1$ = minimum hole diameter, Item 1
$H2$ = minimum hole diameter, Item 2
F = maximum screw diameter

Notes

1. Consideration should always be given to reducing the variety of hole sizes in any one part in order to simplify the fabrication of the part.
2. In *true position tolerancing*, for a *fixed fastener condition* (Item 2 will have no hole size tolerance, that is, Item 2 is a threaded hole):

$$Zt = H1\text{-}F$$

For example, if Item 2 is a threaded hole with a true position tolerance of 0.010 inch, given a minimum hole size in Item 1 of 0.218, and a maximum thread size is 0.190, then:

$$Zt = 0.218 - 0.190 = 0.028 \text{ inch}$$

As $Z2 = 0.010$, then $Z1 = Zt - 0.010 = 0.018$ inch

For a *floating fastener condition* (both Items 1 and 2 have hole size tolerances):

$$Zt = Z1 + Z2 = H1 + H2 - 2F$$

and in the above example, $H1 + H2 = 0.218 + 0.218$, and $2F = 2 \times 0.190 = 0.380$; thus $Zt = 0.436 - 0.380 = 0.056$ inch. If $Z2 = 0.010$, then $Z1 = Zt - Z2 = 0.056 - 0.010 = 0.046$ inch. Thus, you can see that a floating fastener condition allows more tolerance in position.

3. In *coordinate tolerancing*, for a *fixed fastener condition* (Item 2 has no hole size tolerance):

   - Simple linear $Z1 + Z2 = H1 - F$
   - Complex linear $Z1 + Z2 = 1/2\ (H1 - F)$
   - Simple coordinate $Z1 + Z2 = 1/1.4\ (H1 - F)$
   - Complex coordinate $Z1 + Z2 = 1/2.8\ (H1 - F)$

For example, if Item 2 has threaded holes in a complex coordinate pattern, Item 1 has clearance holes with minimum diameter of 0.256 inch, and a maximum thread size of hole size is 0.190, then:

$$Z1 + Z2 = 1/2.8\left(H1 - F\right) = 1.2.8\left(0.256 - 0.190\right)$$
$$= 0.023 \text{ inch}\left(\text{total tolerance for Items 1 and 2}\right)$$

For a *floating fastener condition* (both Items 1 and 2 have hole size tolerances):

- Simple linear $Z1 + Z2 = H1 + H2 - 2F$
- Complex linear $Z1 + Z2 = 1/2\ (H1 + H2 - 2F)$
- Simple coordinate $Z1 + Z2 = 1/1.4\ (H1 + H2 - 2F)$
- Complex coordinate $Z1 + Z2 = 1/2.8\ (H1 + H2 - 2F)$

In the above example, but this time a *floating fastener condition*, then:

$$Z1 + Z2 = 1/2.8\left(H1 + H2 - 2F\right) = 1/2.8\left(0.256 + 0.256 - 0.380\right)$$
$$= 0.047 \text{ inch}\left(\text{total tolerance of Items 1 and 2}\right).$$

Thus, you can see that a floating fastener condition allows more tolerance in position.

References

1. Shipulski M, Six lessons learned from a successful design for assembly program. Hypertherm, Inc., 2006 DFMA Forum.
2. Huthwaite B, Product design for manufacture & assembly (DFMA). Troy Engineering, 1989, International Conference on Industrial and Engineering Applications of A.I. & Expert Systems.
3. Ratay D, Designing microelectronics products for serviceability. Honeywell Inc., 1985 Design News.

Chapter 7
Product Environments

The products the EPE Designer will design must survive the environments that the products will placed in. That environment could be fairly benign, such as a home or office, or it could be rather hostile, such as needing to function in an off-road vehicle or survive underwater.

This chapter will concern itself with the environments that the product is specified to perform in for its customers. So, we'll start by taking a look at some "standardized" environments. Earlier, we have stated that "successful" design can only be defined by the design's ability to conform (or exceed) the overall product specification. The overall product specifications will include a description of the testing required to assure that our product will indeed function in a particular environment.

For example, if our product is expected to function in an office, we'll need to determine what the temperature extremes of the "normal" office are. We will search the literature and existing internal company data to determine what those temperature extremes are. Perhaps there already exists an "industry standard" that will list what the "normal" office temperatures are. Would it be the same for an office in San Francisco, California, as it would for Cairo, Egypt?

After we look at standards for the description of our environments, we'll look at the testing needed to "certify" that we have indeed passed that testing.

Then, in this chapter, we'll move on to describing some common environments such as temperature and vibration, to more fully understand the considerations of these environments.

© Springer International Publishing AG, part of Springer Nature 2019
T. Serksnis, *Designing Electronic Product Enclosures*,
https://doi.org/10.1007/978-3-319-69395-8_7

7.1   Standards

Products exist in a number of different environments, such as:

- Military
- Commercial
- Space
- Transportation

I am most familiar with the commercial environments, and this chapter will mainly address that region. Much of the information will be applicable to other specific environments, so the reader will need to research those specifics if designing into them. This will include products sold into domestic (USA) markets and world markets. Any product sold into a particular locality must pass regulations required in that locality. A host of products are intended for sale into worldwide markets, and research must be done to determine the regulations in all markets. Fortunately, there has been some unification with regulations where many countries recognize the same testing and certification standards. Two examples of "regionalization" of standards are for EMC (see Chap. 9), and for safety (see Chap. 10).

Commercial products can further be parsed as:

- Office (or business)
- Home (or personal)
- Recreational (or outdoors)
- Toys

Of the above commercial listings, I am most familiar with products that go into the survey, agriculture, transportation, and office environments, so a large portion of the commentary will be in those regions. These are mainly products intended for outdoor usage ("weatherproof") but share some commonality with office products in that they contain microprocessors, have storage media, and Internet connections. Consumer products that have PCBAs in them are generally considered "electronic products," and this text is mainly concerning those products as they have a lot of "common ground" as far as design environments are considered.

Product specifications generally list environments where the consumers will use those products. Thus, these products must pass a testing program designed to assess whether or not the product can be used in those environments. As with most engineering endeavors, there already exists many "standard environments" and testing programs designed to determine whether the products pass those tests. Most environmental levels and test programs are available either as:

1. Published industry standards (available for free or at a cost)
2. Company (internal) standards (only available as that company sees fit to distribute them)

Other level and test programs are developed to suit customer needs and the product at hand.

The US defense standard, often called a military standard, "MIL-STD," "MIL-SPEC," or (informally) "MilSpecs," is used to help achieve standardization objectives by the US Department of Defense and is an oft-used source of references for environmental testing (Ref. [1]).

Several examples of how product environments are defined are listed below:

Example 1 Product is to be placed in an automobile, and the customer is the automobile manufacturer (OEM). The automobile manufacturer will have very specific specifications on the environmental levels and testing needed. These will likely include existing standards from the customer (per Ref. [2]):

- ASTM B117 (American Society for Testing and Materials)
- USCAR-1   (US Council for Automotive Research)
- IEC 60529   (Int'l Electrotechnical Commission)
- GMW 3172 and 16,390 (General Motors Corporation)
- ISO 16750-4   (Int'l Organization for Standards)
- CETP (Ford Corporation)
- SAE (Society of Automotive Engineers)

   Testing would include:

- Environmental exposure
- Thermal Shock
- Shock pulses
- Chemicals testing (for automobiles)
- Durability testing
- Tensile strength
- Dust-settling method
- Salt spray/salt fog
- Rain simulation/submersion
- Ingress protection
- Noise measurements
- Battery testing
- Altitude simulation
- Vibration testing
- Failure Mode Verification Testing (FMVT)
- High-pressure spray
- Buzz, squeak and rattle
- Electrical testing

   So, for Example 1, a starting point would be to check internal resources (marketing, previous products, other groups within the company) to see if this is a complete set of environmental specifications. Then, consult those test specifications to understand their appropriateness. Then, a decision on whether that testing will be done in-house, or at an external laboratory (contracted testing) needs to be determined.

Example 2 A product will be designed to be used by a surveyor. The use case is to mount the product on the top of a 2 meter length pole. As it is "reasonable" in standard customer usage for the pole to fall over (and thus, the product gets exposed to a 2 meter drop), the question becomes what (standard) testing is necessary to test for this "environment?" After a literature test on this type of testing, it seems that no "industry standard" test exists. A review of "like products" sold by the competition doesn't give any details on their drop test environment (other than the product "passes" a 2 meter drop). Therefore, internally, the product design team must decide exactly what is meant by a "2 meter fall." Possible testing could be proposed as:

- Product to still work after passing "x" drops (say, ten drops)

   Questions need to be considered such as:

- What is the definition of "still work?"
- What cosmetic damage is permitted?
- Can a replaceable "piece" be added (after the fall) that will be sacrificed for the event? (Later to be replaced by the customer)
- Is the "bounce" that happens after a fall controlled (will it be a "repeatable" drop)?
- Does the product and pole get dropped on concrete, asphalt, or dirt? (And what is the "definition" of, say, concrete?)
- Should the ten drops be controlled so that the object gets hit with a 36° (360/10) rotation so that the drops occur *evenly* spaced on unit? (Or are all ten drops hitting at the same spot, or a random spot, on the unit?)
- Is the product turned on (functioning), or is it powered down when the drops occur?
- At what temperature are the drops to take place (ambient, high or low temperature extremes)?

These and many more questions could be asked when going about defining this "drop environment." The designer cannot design the product without addressing what is required by the specification with regard to "passing" this drop test.

So, for Example 2, where no standard test (apparently) exists, it must be carefully considered, ahead of the design process, as to what testing the product must pass. Some early prototype testing may provide some answers as the design develops.

This chapter on product environments will contain sections on vibration, shock, temperature, moisture, and other considerations. EMI control (EMC) will be addressed separately in Chap. 9. Safety agency certification will be addressed in Chap. 10.

7.2 Testing

We have been following a path in the design process. This path generally looks like:

- Specification – the definition of *what* we want to design
- Project management – the *plan* to deliver the *what* (time/budget/personnel)
- Testing program to determine if design meets specification
- Delivery to customer

So, the testing methodology is usually included in both the specification and project management plans. All test specifications and testing plans are an approximation to the environment that the product will operate (and survive) in. That is, if it is "stated" that a particular consumer product is to be exposed to a maximum temperature of 60 °C, this, in itself, raises a few questions:

1. Should we test at higher temperatures than 60 °C to see if we have "margin?" If so, how much higher?
2. If we test at 60 °C, how long should the test be so that we say it "passes?"
3. Is there some "equivalent testing" that we can do to reduce the testing time? That is, is testing at 65 °C for 1 hour, equivalent to 60 °C for 10 hours?
4. If we are to continue testing this unit (say, for vibration testing), is it acceptable to "stress" the test unit for *both* the temperature testing *and* the vibration testing? Which testing should come first (temperature or vibration), that is, what is the "testing order?"
5. How many units should we test for the test to be considered passing? Is one (of one) unit passing considered a "pass," or must it be, for example, ten of ten? If we decide that we will test ten units, what should we do if nine units pass and one unit fails the testing?
6. Is the testing that we do going to be shared with either the customer or agencies that will use the results to determine whether we pass their definition of the testing? How much test documentation is required? Where will the testing results be stored?
7. Will the testing be done with in-house resources or contracted out to an external facility? If the testing is contracted out, how much "pretesting" is to be done before we consider the test units "worthy" of testing by that external facility?
8. If the unit is composed of PCBA, cables, and an enclosure, should we do separate testing on the PCBA, separate testing on the cables, *and* the assembled version of PCBA + cables + enclosure? The product specification would only state that the assembled version should pass testing; however, it may be prudent to "pretest" the individual components to aid in failure analysis.
9. How close is the unit we are testing to the unit that the customer will see? Is the testing done on a "prototype," preproduction, or production unit? Will we continue to test production units to make sure that the product specification is being met at further customer deliveries?

These and other questions exist for any test program, and it is up to each corporation and design team to answer those questions.

Many standard industry testing programs exist to help determine both the limits of testing and the testing methods. One such test method is HALT (Highly Accelerated Life Testing), and it is usually applied to solid-state electronics to determine failure modes, operational limits, and destruct limits. HALT applies one stress source at a time to a product at elevated levels to determine the levels at which the product stops functioning but is not destroyed (operational limit), the levels at which the product is destroyed (destruct limit), and what failure modes caused the destruction of the product. HALT typically uses three stress sources: temperature, vibration, and electrical power. Each stress source is applied starting at some nominal level (e.g., 30 °C) and is then elevated in increments until the product stops functioning. The product is then brought back to the nominal conditions to see if the product is functional. If the product is still functional, then the level at which the product stopped functioning is labeled the "operational limit." The product is then subjected to levels of stress above the operational limit, returning to nominal levels each time until the product does not function. The maximum level the product experienced before failing to operate at nominal conditions is labeled the "destruct limit." This process is repeated for hot temperature, cold temperature, temperature ramp rate, vibration, and voltage. The process is also repeated for combined stress.

Some testing programs use "accelerated" (higher) levels of the environment (temperature, vibration, etc.) at shorter test time lengths to gather predictive data as to how long a product will last at the "worst-case" levels.

In addition to testing programs, quality control plans should be developed. A common procedure to implement such a plan is a FMEA (Failure Mode and Effects Analysis) where the following are identified:

- Potential failure modes
- Severity of failure
- Cause of failure
- Probability of occurrence of failure
- Alternative designs to mitigate failure

I'd like to make another note about the general test plans (and planning). There are several "stages" or types of testing which I'd like to list below:

1. "Quick" testing just to prove (or disprove) a concept under consideration. This testing's emphasis is on "information gathering" and may be quite inadequate on documentation. Again, speed is the main driving force. However, care must be taken to access the information's appropriateness to the design goals and not be too quick to throw away good ideas (or keep under consideration "bad" ideas).
2. Testing that is well-documented and will serve to prove that the design did or came close to passing the specification. This documentation may be the only residual of why a particular design option was chosen. These testing programs will also serve to prove that a particular specification was "passed."
3. "Certified" testing either done in-house or contracted out. This testing is done to formal industry standards and procedures. They are signed documents that prove

that a particular test has been passed. If a government agency requires the testing and passing of a test, this documentation will provide proof that the design has passed.

7.3 Temperature

An example of an environment that is tested for is Operating Temperature Range Cycling. This test is to determine if mechanical expansion and contraction cause component malfunction, deterioration, or failure (e.g., potting cracks or lead breakage). The test samples are cycled between the temperature extremes defined below.

- 7.3.1 The nominal operating temperature requirement for the components is −40 °C to +80 °C unless otherwise specified. This allows the use of industrial-grade components without ignoring internal heat rise.
- 7.3.2 The thermal cycling specified in the test constitutes an aging process and as such should be performed prior to any other environmental testing. The device under test (DUT) should be checked for normal operation throughout the operating temperature range.
- 7.3.3 Follow the two steps as outlined below. The minimum dwell time at the temperature extremes is 1 hour; this is required for the unit to reach thermal equilibrium. When moving from one temperature extreme to another, the temperature should be ramped at a minimum rate of 5 °C/min. After changing the input voltage, cycle power to verify that the unit is booting properly.

  Step 1: Air temperatures in the environmental chamber shall be measured to ensure compliance with the specified limits and rate of change.

  Step 2: Power shall not be applied to the component throughout the test. A functional test shall be performed at least once in every five thermal cycles throughout the operating voltage range, except at the lower storage temperature (usually −50 °C) where the component shall not be operated. It is desirable to perform this functional test during the temperature transition in the event broken leads may not be making contact due to differences in thermal expansion rates.

- 7.3.4 Note: If any unit fails at 80 °C, reduce the max temperature by 5 °C until all cycles pass. Note the maximum temperature that all units pass.

  Dwell times were 1 hour minimum.
  There were 20 cycles that generally looked like:

- Dwell 1 25 °C, Ramp 1
- Dwell 2 −40 °C, Ramp 2
- Dwell 3 80 °C, Ramp 3
- Dwell 4 −40 °C, Ramp 4
- Dwell 5 80 °C, Ramp 5
- All the way to Dwell 42 25 °C

Pass/fail criteria:

- The test samples shall function within specifications.
- Visible cracks or damage are not allowed.
- Once the remaining tests are complete, disassemble the test samples, and inspect for evidence of corrosion or moisture entry. Internal moisture is not allowed.

Above Operating Temperature Range Cycling was followed by thermal shock testing:

- 7.3.5 Place the DUT into a temperature chamber, set to 25 °C, and allow the unit's temperature to stabilize.
- 7.3.6 Set the input voltage to 24 V, and verify that the unit is operating normally.
- 7.3.7 Power down the DUT.
- 7.3.8 Subject the unit to 10 temperature cycles with a minimum temperature of −55 °C and a maximum temperature of 85 °C. Dwell time at the temperature extremes should be 1 hour minimum. The ramp rate should be as fast as the chamber will allow.
- 7.3.9 Set the chamber's temperature to 25 °C, and allow the unit's temperature to stabilize.
- 7.3.10 Repeat 7.3.6.
- 7.3.11 After completion of the thermal shock test, subject the component to the vibration levels specified in vibration/shock/drop for 10 min. in each of the three mutually perpendicular planes. Check its functionality while it is being vibrated. No failures are allowed.

(Same pass/fail criteria as for the Operating Temperature Range Cycling Test)

Again, the above temperature testing is just an example of one company's methodology toward creating a standard test.

Some "worst-case" use environments for surface-mount electronics are shown in the Table (see Ref. [3]) just to show that different use categories have different levels of environments:

Use category	Normal use T_{max}	Years of expected use (approx.)
Consumer	60 °C	1–3
Computers	60 °C	5
Telecomm	85 °C	7–20
Commercial aircraft	95 °C	20
Military ground/ship	95 °C	5
Space	85 °C	5–20
Military avionics	95 °C	10
Automotive passenger	95 °C	5
Automotive under hood	125 °C	5
Industrial	95 °C	10

7.4 Vibration

Vibration testing is a very common test that most products must pass. Vibration (as a test method) is applied to a solid-state electronic product for two reasons:

1. Vibration (some small level) will induce a failure very quickly, and that tests the basic design of items such as PCBA support, interconnects between that PCBA to other internal assemblies, and overall assembly integrity.
2. The product will be exposed in its usage to at least that vibration level.

All products undergo some levels of vibration because, even if they will perform in a benign environment (such as an office), they still will be exposed to at least the shipping environment (from assembly line to customer). Thus, the "shipping container" becomes a key element in the (overall) operation of the electronic product, and this is further explored in Chap. 11, Shipping and Packaging.

Typical defects detected by vibration testing include intermittent solder connections and printed circuit board traces. There can also be loose or broken components that were not discovered during visual inspection or performance testing. Unless these defects are found, they tend to cause failures early in the product life. Early failures are known as infant mortality failures and are illustrated by a classical bathtub-shaped failure rate curve (aka, Weibull curve). See Fig. 7.1 and Ref. [4].

For many years, typical vibration tests started at one frequency and swept through some frequency range, shaking the unit at a single frequency at a time. This technique is still widely used because it is simple and requires relatively inexpensive instrumentation. But there is a problem with these simple vibration tests: they can only excite a single resonance at a time in the assembly under test. Such tests do not detect any mutual influence of resonances that may exist at different frequencies. However, a technique called random vibration will show up problems caused by interrelated resonances. Random excitation vibrates the unit under test at all frequencies in a given bandwidth simultaneously (typically 20–2000 Hz). Random

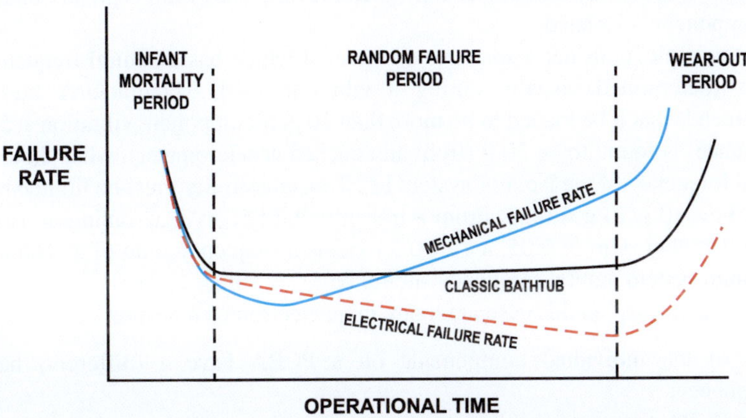

Fig. 7.1 Bathtub curve

vibration represents the real world, one in which all frequencies are present at the same time in varying amplitudes. This is what the product will experience in both transportation and actual usage.

A tremendous reference in the area of vibration testing is Ref. [5]. Steinberg's text serves as a "constant companion" for the basics of good design and modeling of the vibration environment. Vibration is a complicated subject, and this text can only introduce some of the considerations.

As structures should be designed so that they are not exposed to vibration environments at the structure's natural frequency, a basic tenet of the design is to:

1. Find out through analysis (verifying by experiment) the structure's natural frequency
2. Determine if that structure will be exposed to vibrations (at what levels and frequencies)

The first step in a vibration protection program is to list the natural frequencies for the damage-prone components. This is done by calculation or by testing the product on a shake table. In the test, components are illuminated by a strobe light flashing 1 or 2 Hz out of synchronization with the table frequency. Testing with a strobe light is good for frequencies up to about 75 Hz, where movement becomes too small to be seen. Above 75 Hz, either calculations or accelerometer measurements will be needed to determine natural frequencies.

If the whole system is to be isolated on shock mounts, resonant frequencies of the individual components must be avoided. If the system natural frequency matches the component natural frequency, the component will resonate excessively and malfunction. Therefore, choosing a system natural frequency means identifying component frequencies and selecting a system frequency between them (or, below all of them) to isolate all components. A transmissibility diagram shows the relationship between input forces and responding vibrations. Transmissibility, T, the ratio of transmitted force (or acceleration) to disturbing force, is plotted against the ratio of disturbing frequency, f_d, to natural frequency, f_n. When the frequency ratio is 1, resonance, or vigorous oscillation, occurs. When the frequency ratio is greater than $\sqrt{2}$, the component is isolated.

For example, consider a sensitive component which has a natural frequency of 50 Hz (determined visually during a vibration test). Manufacturer literature recommends that it be loaded to no more than 10 g. If component vibration at 50 Hz (resonance) is found to be 30 g (from an attached accelerometer), what should the natural frequency of the isolator system be? Transmissibility must be limited to $T =$ Fout/Fin = 10 g/30 g = 0.33. From a transmissibility curve, assuming an isolator with a damping ratio of 0.05, a T of 0.33 yields a frequency ratio of 2. Therefore, maximum system natural frequency should be $f_\mathrm{d}/2 = 50/2 = 25$ Hz.

The "structures" in this case are many in an electronic enclosure:

- All of the individual components on a PCBA have a (different) natural frequency:
- The PCBA (as an assembly, i.e., components plus printed circuit board) has a natural frequency.

- Electronic Box Assembly (multiple PCBA plus other assemblies, plus enclosure) has a natural frequency.

So, the problem with vibration analysis is that a vibration input (level and frequency) that is applied to the Electronic Box Assembly may directly "pass through" to the PCBA (or to the individual components on the PCBA). "Pass through" meaning that the PCBA may see the *same* vibration input. Or, that vibration (level and frequency) may be amplified or damped by the component mounting system.

The following is a checklist for determining the vibration problems with the design:

1. In the early stages of design, determine or estimate the natural frequencies of the sensitive components. Support these components as firmly as possible. For example, locate a sensitive individual PCB component near the PCB mounting holes.
2. In the preliminary design stage, order a range of system isolators, in terms of size, stiffness, and type. Determine system natural frequency, and select a frequency lower than that calculated or measured for the troublesome components. Perform vibration tests with simulated components and masses on real isolation mounts.
3. Run random, swept sine, and shock pulse of different shapes (see next section), drop, and shock response tests with real components to determine how the design is functioning.

7.5 Shock

Single impacts differ from continuous vibration in that they are usually of larger amplitude, of short duration, and cause a different structural response in the product. That is, a single shock pulse will cause components of many different natural frequencies to respond, or ring, but with varying amplitudes. This is called the shock response spectrum and is illustrated by shock transmissibility curves. These predict how large a response some components with a known natural frequency will have to an input of a certain pulse length. The key to understanding the shock transmissibility curve is recognizing the effect of the ratio of input shock pulse period to natural frequency.

Basically, input acceleration is absorbed by a resilient mount, and the shock energy is released over a broader time base. By dispersing the shock energy over a broader time base, the output accelerations are reduced. This can be illustrated by the two drops shown in Fig. 7.2:

- Curve 1 with a Higher *G* Input over a shorter time interval
- Curve 2, showing the effect of the addition of a shock isolator, resulting in a Lower *G* Input spread out over a longer time interval. Note that the *total* energy dissipated is the same in both drops

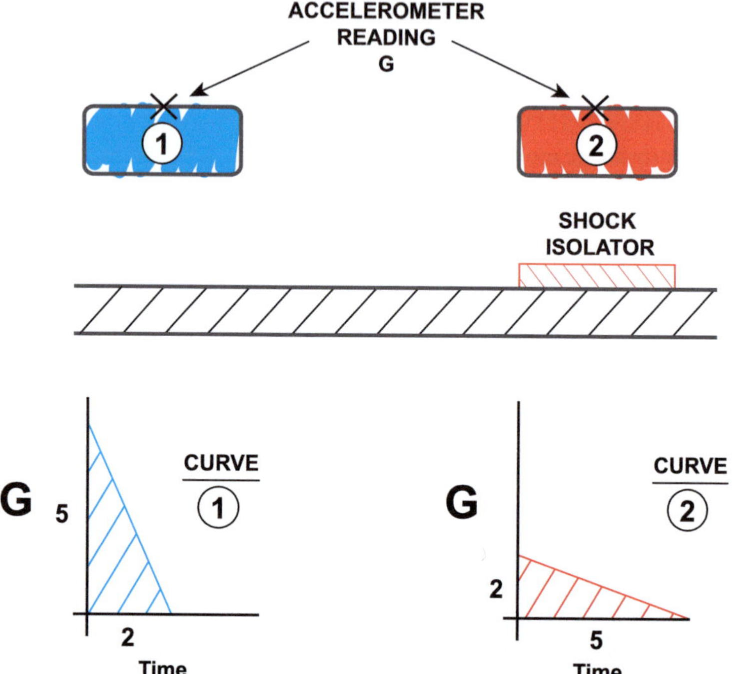

Fig. 7.2 Shock isolation

This would be the goal of a shock isolation design. The designer must also accommodate sway space within the product design. If not, even though the shock isolator may be very efficient, lack of necessary sway space may cause secondary collisions resulting in the same damaging effect as if no shock attenuation devices were used.

For example, an automobile with a natural suspension frequency of 3 Hz hitting a short bump causes almost no reaction inside the car. However, if it hits a bump that extends for a few feet, the car may lurch violently upward. This illustrates the effect of the ratio of impact duration to system natural frequency: if small, the reaction is brief; if large, the reaction is large. Let's further say that:

If the shock input was measured as 500 g at 0.5 ms, what effect would this have on a 50 Hz natural frequency component?

First of all, shock transmissibility curves exist for various pulse wave shapes and various system damping ratios. Shock transmissibility is a plot of "Amplification Ratio" ($Gout/Gin$) vs. "Frequency Ratio" ($fn/fpulse$).

$$f\text{pulse} = 1/2t = 1/(2 \times 0.0005) = 1000 \text{ Hz.}$$
$$fn/f\text{pulse} = 50/1000 = 0.05$$

The shock period ratio would be $t/fn = 0.5 \times 10^{-3}$ sec $/ (1/50$ sec$) = 0.025$

From a shock transmissibility curve for these conditions (saw-tooth shock pulse, zero damping):

$$Gout = fn \,/\, fpulse \times Gin$$
$$Gout = 0.05 \times 500 \text{ g} = 25 \text{ g}$$

If this would be too high of a response amplitude, then a softer isolation system that has lower natural frequency (*fn* < 50 Hz) would be needed.

As stated in Ref. [2], "Shock analysis techniques can become quite complex, unless some simplifying approximations are made." The author goes on to state some fixes for a PCB to survive high shock levels:

Raise the PCB resonant frequency by:

1. Increasing the PCB thickness
2. Adding more copper planes in the PCB, which increases stiffness
3. Using polyimide glass that has a higher modulus of elasticity than epoxy glass
4. Adding a stiffening rib if space is available on the PCB
5. Moving the larger components away from the (unsupported) center of the PCB toward the supported edges

A great example of an evaluation of a system with isolators is shown in Ref. [6]. A set of isolators must be selected that will work with both the vibration and shock environments. This is shown to be done by trial and error, where an isolator resonant frequency is assumed and then evaluated for both environments. If the selection is poor (dynamic single-amplitude displacement, "sway," is too large), another selection is made, and the process is repeated until an acceptable system is obtained. This example also checks to see if the bending displacement of the PCB with isolators is acceptable.

Another example (as shown in Ref. [7]) is a good example of a "typical shock problem." It starts with shock fragility and equipment weight as the starting points. If the fragility of the equipment in a shock environment is the critical requirement of the application, the natural frequency of the system will depend on the required isolation of the shock input. Also, keep in mind that the *system natural frequency under a shock condition will typically be different from that under a vibration condition* for systems using elastomeric vibration isolators.

Fragility is defined as the amount of vibration or shock which a piece of equipment can take without malfunctioning or breaking. In isolation systems, this is a statement of the amount of dynamic excitation which the isolator can transmit to the isolated equipment.

Step 1: Determine the shock input in terms of (either):

$$\text{Drop Heights}, V = \sqrt{2gh} \, (\text{inelastic impact})$$
$$V = 2\sqrt{2gh} \, (\text{elastic impact})$$
$$\text{half sine acceleration}, V = \left(2g/\pi\right) \text{Ao } t', \text{ where}$$

Ao is the shock pulse max and t' is the time interval.

For this example let's pick a 15 g half-sine, 11 ms shock pulse, so that

$$V = \left(2g/\pi\right)\mathrm{Ao\,t'} = \left(2\times386\,\mathrm{in/sec^2}/\pi\right)15\times0.011\,\mathrm{sec} = 40.5\,\mathrm{in/sec}$$

Step 2: Estimate the fragility factor of the equipment and express in terms of g's. This information is normally obtained from component manufacturers or test data.

For this example lets pick $G = 8$ g.

Step 3: Determine the natural frequency, fn, required to attenuate the shock input levels at or below the fragility factor, as

$$fn = \left(\mathrm{g}/2\pi\right)\left(G/V\right) = \left(386/2\pi\right)\left(8/40.5\right) = 12.1\,\mathrm{Hz}$$

Step 4: Determine the isolator deflection, Δd (inch), and sway space required by $\Delta d = V/(2\pi\,fn) = 40.5/(2\pi\,12.1) = 0.53$ inch

Step 5: Calculate the load at each mounting point. If the center of gravity is at the geometric center, simply divide the total weight by the number of mounting points. If the load is eccentric, perform a basic load distribution analysis.

Step 6: Choose an isolator (from an isolator catalog) that matches the fn and load range.

7.6   Water Protection

7.6.1   Standards Available

Most office products need some protection from, at least, the water in the air (humidity). There are humidity testing standards and products must be designed for these conditions. Products that must survive the harsher effects of outdoor environments are faced with designing some sort of sealing integrity. For example, if one is to design a cell phone, the designer faces some obvious questions (with regard to moisture):

- Will the design protect against rain? (How is "rain" defined?)
- Will the design protect against immersion into a puddle of water (or, perhaps a toilet)? (How deep is the submersion?)
- What about the effects of a heating/wetting cycle where moisture may be pulled in due to an internal vacuum?
- Can the consumer get injured when the equipment is exposed to this moisture?
- What are the warranty implications of a seal failure?
- Do we test every unit that is built on the assembly line for sealing integrity?
- Is condensation (fogging) an issue on the inside of the unit?

The waterproofing issue is a complex one. It's a costly process involving the basic design of electronic products and the cases that house them. Then there's the matter of determining how moistureproof a given piece of equipment should be and how to verify it. *None* of the popular "water-integrity" terms (e.g., *weatherproof, splashproof, water-resistant*, even *waterproof*) have real definitions. The only universally accepted truth in the issue of waterproofing is "moisture and circuit boards don't mix."

Fortunately, the electronics industry seems to have embraced the IEC 529 Protection Levels for protection against contact with live (powered) parts and the ingress of foreign bodies, including:

1. A person touching live parts with either with their hands or tools
2. Penetration of foreign objects (solid and dust) into the equipment
3. Degree of water protection (dripping, splashing, jets, and submersion)

Other standards certainly exist that may be more particular to either the specific type of equipment or particular region of the world. Examples of which include:

- NEMA (National Electrical Manufacturer's Association)
- DIN 40050 ("German Institute of Standardization")
- JIS-3 (Japan Industrial Standard)
- MIL-STD-810, Method 512.4 Immersion
- ANSI/IEC/EN 60529

But, they have similar attributes of IEC 529. Again, a design specification would specify the standards for water protection appropriate to their product and their customer locations.

7.6.2   IPXX

The IEC 529 Protection Levels (protection against water), or IP Rating (Ingress Protection Rating) range from a rating of "0" (non-protected) up to "8" (protected against submersion). A very common rating for electronic products would be IP67 where the (first characteristic numeral) "6" is dust-tight protection against live parts and foreign bodies, and the (second characteristic numeral) "7" is protection from the effects of immersion in water.

An IP67 test (for the second characteristic numeral 7, see Ref. [7]) is made by completely immersing the enclosure in water in its service position so that (basically) the IUT is 3.3 feet below the surface of the water for duration of 30 min.

Another common test is IP65 where the "5" is protection against water jets (heavy rain). This test is (basically) spraying water on the item under test with:

- A nozzle that is 0.25 inch in diameter, at 3.3 GPM
- 8 foot distance (nozzle to enclosure surface)
- 3 min minimum test duration

7.6.3   Sealing

Beyond the particular specification to be met, the designer is faced with how to proceed with the design. Of high importance will be:

- How to design a seal around *any* opening of the box. Usually, there is a "main seal" between an upper and lower portion of the box. This opening basically allows insertion of the electronics into the box probably during an assembly process. Other openings result from interfaces to the equipment user such as:

  a) Connectors or cables
  b) Visual devices such as LCDs (touch screens) or LEDs
  c) Audio devices
  d) Input interfaces such as buttons or pads
  e) Media or memory devices such as disk drives, card readers, and memory sticks
  f) Areas around fasteners

  Most of these openings are in the category of "input" or "output" to/from the user.

  Remember, any opening to the box will be a potential area of leakage. Problems must be solved such as:

- The design must seal under *all* of the environmental conditions:

  a) Temperature excursions
  b) Vibration and shock occurrences
  c) Moisture in the air that can get through seals
  d) UV radiation that can attack seal integrity
  e) Chemicals that can attack seal integrity
  f) Creep effect where material properties change with time under loading
  g) Change in altitude (external pressure change)

- How to test for any leaks and identify quickly exactly which opening(s) is leaking, during the assembly process.
- Should a port be added to the electronic enclosure whose sole purpose is to allow leak testing (and will be sealed 100% after a successful leak test)?
- If repairs are made in the field (servicing), how will repairs be undertaken to maintain seal integrity? How will the repaired unit be tested for sealing?

7.6.4   Water Vapor

I'd like to list some of my own observations on a chapter from Ref. [8], again offering some quotes from the author. This is from Sect. 2.9 (of Ref. [8]) Sealed Electronic Boxes:

- Slight amounts of condensation on sensitive components, PCBA, or electrical connectors produce large changes in operating characteristics of the system; therefore to minimize potential problems resulting from humidity, moisture, and condensation, these systems are often packaged in sealed electronic boxes.
- Many different types of seals can be used, depending upon the size of the unit, the cost, and the ease of repair. Sealing types can be divided into two main types of sealing, either "face" or "piston." O-rings can be used as either type of seal, by sealing:

  a. Between a flange and cover (face seal)
  b. Between inner and outer circular geometry (piston seal)

- Pressure relief may be necessary from the sealed (internal) volume and the outside environment. This may be due to either:

  a. High internal pressure due to increase in altitude environment.
  b. To reduce "pumping" action between outside air and inside air. This is normally accomplished with a "GORE-TEX patch" (a small piece of "breathable" material) which allows air to exit or enter the box but prevents water vapor (in the air) from entering the box.

- Flat elastomeric gaskets are easy to use and cheap to fabricate. These gaskets are capable of providing an airtight seal for a large electronic box. This type of seal should not be used for boxes that require a water vapor seal. The interface pressure between the flat gasket and its mating box and covers is usually very low, because the elastomers are free to deform as the cover screws are tightened. Although this gasket will seal in air, it will not seal out water vapor. The water vapor molecule is smaller than the air molecule, so that the water vapor molecule will pass through restricted openings more readily than the air molecule.

The water vapor molecule is smaller than the air molecule because the water vapor molecule has a smaller molecular weight. The size of the gas molecule is related to the molecular weight of the gas, so that a gas molecule with a high molecular weight will be larger than a gas with a low molecular weight. Air is primarily a mixture of:

- Oxygen (O_2) with a molecular weight of 32
- Nitrogen (N_2) with a molecular weight of 28

  Water vapor is still water (H_2O):
  Two atoms of hydrogen (H_2) have an atomic weight of 2.
  One atom of oxygen has an atomic weight of 16, for a total molecular weight of 18.

  The smaller size of the water vapor molecule therefore permits it to pass through openings that would normally restrict the flow of air molecules. The O-ring type of gasket seal has a very high interface pressure, which makes it very effective for boxes that require a water vapor seal.

Water vapor is a gas and therefore obeys the normal gas laws. The force driving the water vapor migration is the difference in the partial pressure of the water vapor between two different areas. Water vapor will tend to flow from the high-partial-pressure areas to the low-partial-pressure areas. If an electronic box is assembled in a dry climate, the partial pressure of the water vapor within the box will be low. If that box is now transported to a humid area, such as Florida, the partial pressure of the water vapor in the atmosphere will be quite high. The water vapor will try to equalize the partial pressure, so that it will try to enter the box. It does not matter if the absolute air pressure within the box is 50 psia and the outside ambient pressure is only 14.7 psia. Since the partial pressure of the water vapor within the box is low, the tendency will be for the water vapor to try to enter the box to equalize the partial pressures.

If the type of gasket used to seal an electronic box is not effective, water vapor will pass through the seal and condense within the electronic box. It does not take very long for half a pint of water to accumulate within a small box due to condensation. An electronic box 15 inch deep × 10 inches wide × 8 inches high, with a flat rubber gasket seal under the top cover, can accumulate that amount of condensation in a period of only about 3 months when it is continually exposed to humid air.

The basic structure of the electronic box itself may not be capable of providing an effective water vapor seal. Sheet metal structures are generally satisfactory; however, cast structures are often very porous. Some magnesium and aluminum castings 1 inch thick will not hold air for more than 5 min. Structures of this type can often be impregnated with an epoxy resin, usually applied by means of a vacuum.

Desiccators may be placed within the electronic enclosure to absorb any water vapor that may seep into the sealed box over a long period of time.

7.6.5 *Other Notes on Sealing from Moisture*

1. Another valuable standard test for "real-world" sealing integrity is a "rain and shine" test. This standard test covers the seal integrity of many electrical enclosures and outdoor products. The rain and shine test is a very severe sealing test where the test samples are heated with a heat lamp for a period of time (typically 1 hour) and then subjected to a water spray for an equal period of time. The water spray cools the sample, creating an internal vacuum which draws in moisture through any poorly sealed openings. The expansion and contraction cycles result in a "breathing" effect (heat pumping) on the component, allowing moisture ingression. The heating/wetting cycle is repeated a prescribed number of cycles and the samples inspected for moisture entry. Typically, the test has:

   Simulated sunshine (500 watts of heat, 60 cm above the test sample), 1 hour, followed by
   Simulated rain (5.0 mm of water/hour delivered at a 30° angle from a nozzle with a solid cone spray), 1 hour

100 test cycles of above

2. Some testing (MIL-STD-810) has an immersion test. This testing dunks a pre-heated unit to a meter under water. The *total* pressure (change) is about 1.5 psi from the water depth and another 1.5 psi from the change in temperature (for a total change in pressure of 3.0 psi).

33′ of water = 14.7 psi, and thus 3′ of water = 1.5 psi
ΔP from temperature goes as ideal gas law, $PV = nRT$ (°R), then with
$T1 = 122$ °F and $T2 = 68$ °F
$P2 = ((460 + T1)/(460 + T2)) \times P1 = 1.1 \times 14.7 = 16.2$ psi
$\Delta P = 16.2{-}14.7 = 1.5$ psi

3. Leak Testing Methodology

Various leak detection methods can be used to detect and locate leaks. Ref. [9] lists the "pros and cons" of the various technologies:

- Bubble testing
- Halogen sniffing
- Pressure rise
- Pressure decay
- Helium mass spectrometry

Most electronic product assembly lines would employ either pressure rise or decay. The sensitivity of these methods is about 10^{-4} cc/sec in a production environment. Leak rates can be calculated from the collected data using:

Actual leak rate = ($\Delta P \times$ Internal Volume)/time elapsed during pressure change.

7.7   Other Design Environments

The designer could be designing for just about any condition imaginable. All of these conditions are conditions that the product will survive (and prosper) in – these are the environments that the customer needs. It could be anything from "Martian Terrain" to "pickle juice." Some real environments that are fairly common are:

- Solar radiation
- Acoustic noise
- Oil resistance
- Salt fog
- Corrosion
- Ozone
- X-ray radiation
- Electrostatic discharge (ESD)
- Fungus
- Power supply variability

- Explosive atmosphere
- Gunfire vibration

It will be the designer's task to fully understand the test requirements and test conditions that will be acceptable to the customer.

References

1. United States military standard, Wikipedia
2. Automotive testing website, Intertek Corporation
3. Charles HK Jr, Hoffman EJ (1993) Extreme environments. Advanced Packaging Magazine
4. Bhote AK, World class reliability: using multiple environmental overstress tests to make it happen
5. Steinberg DS, Vibration analysis for electronic equipment. Wiley-Interscience
6. Products to control shock, vibration, & noise. Barry Controls Catalog
7. CEI IEC 529, International Electrotechnical Commission Publication
8. Steinberg DS, Cooling techniques for electronic equipment. Wiley-Interscience Publication
9. DeLuca J, How much can it leak? Alcatel Vacuum Product, Assembly Magazine

Chapter 8
Cooling Techniques

This chapter could have been titled "heat dissipation," "heat transfer," or several other candidates. The EPE Designer will be faced with dissipating heat so that the components of the equipment:

- Keep within their temperature operating limits
- Run as coolly as possible to increase the reliability
- Do not burn or cause undue irritation to the equipment users

All cooling techniques and issues can occur both on the individual component level and, all the way up to, the system level. We could be talking about cooling a single chip on a PCBA or talking about keeping the surface temperature of the device enclosure (with that single chip) cool to the touch.

We will start with some "building blocks" by looking at what is generating the heat loading and critical temperatures within our product. We will move on to describing the three modes of heat transfer that the EPE Designer can use to dissipate heat. The three modes of heat transfer are:

- Conduction – where heat is dissipated by contacting the hot element with a "conductor" and "drawing the heat away."
- Convection – where heat is dissipated by air currents that naturally happen as hot air rises from the hot element which draws the heat away from the hot element and transfers this heat to cooler air above. This happens "naturally" (without any air moving device, such as a fan), or it can happen *with* a fan and then is known as "forced" convection.
- Radiation – where heat is "radiated" (emitted) just by virtue of its temperature. This can happen in a vacuum, that is, air is not needed for the radiation mode of heat transfer. Radiant energy is sometimes envisioned to be transported by electromagnetic waves.

We will discuss in the chapter the relative "strength" of each of the modes of heat transfer and how they are used to cool hot spots in our designs.

© Springer International Publishing AG, part of Springer Nature 2019
T. Serksnis, *Designing Electronic Product Enclosures*,
https://doi.org/10.1007/978-3-319-69395-8_8

We'll conclude with a discussion on cooling techniques to use as the power density (wattage per available volume) goes up, and conduction and convection are inadequate to cool the product.

8.1 Heat Loading and Critical Temperatures

By definition, all electronic equipment requires power (energy) to operate. Power is the rate at which energy is consumed (or generated). When a light bulb with a power rating of 100 watt is turned on for 1 hour, the energy used is 100 watt-hours. For example, in the case of a 100 watt (rated) light bulb, about 10% of the energy is converted to light, while 90% is dissipated as heat. Whenever electrical current flows through a resistive element, heat is generated in that element. An increase in the current or resistance produces an increase in the amount of heat that is generated in the element. If the heat flow path is good, the temperature may rise until it stabilizes at a point where the heat flowing away from the element is equal to the heat generated by the electrical current flowing in the element. Heat always flows from the hot area to the cool area. Since the electronic components are usually the source of the heat, the electronic components will usually be the hottest spots in an electronic system. The basic heat transfer problem in electronic systems is, therefore, the removal of internally generated heat by providing a good heat flow path from the heat sources to an ultimate sink, which is often the surrounding ambient air.

The design engineer has the basic three modes to dissipate heat from the overall design (or singular component): conduction, convection (natural and forced), and radiation. Usually, all three modes of heat transfer are available to dissipate heat, but one method may dominate the design.

It may be interesting to note that there are some "figures of merit" that exist for the "probable" solution choices between conduction, convection, and radiation (as the main mode of cooling). For example, the term "thermal density" can refer to heat generated per unit volume (or unit area). As the thermal densities increase (from 0 to 10 watts/liter volume), the main choices for cooling are:

1. Natural convection air cooling
2. Forced-air (convection) cooling
3. Liquid cooling (heat exchanger, immersion in liquid, or other "exotic" cooling methodology)

Note that the general cost and complexity increases as we go away from the use of natural convection as the solution.

Generally, if the thermal density is below 10 watts/liter (0.16 watts/in^3), natural convection is a viable solution. Between 10 and 100 watts/liter (1.6 watts/in^3), forced-air cooling is needed, beyond which some sort of liquid cooling may be required.

Most electronics are sensitive to heat – and excessive heat (or cold) can impair reliability of the equipment. It is very important to implement a good thermal design for electronic equipment because there are cases in which the whole system can be damaged by a malfunction or fault in a single component.

The most important steps in determining what method will be used to dissipate the heat loading are:

1. Determining the input heat loading and/or maximum surface or substrate temperatures permitted. Problems usually start with known heat loading and with maximum component surface (or junction) temperatures known.
2. Determining whether the electronic enclosure needs to be completely sealed (environmentally). If the box can be vented to allow the inflow and outflow of air, then convection methods (natural or forced) can cool the components. If the box is sealed, no holes for the passage of air can exist. Sometimes, sealed enclosures are cooled by additional (closed) systems that can take away the heat from hotter areas. A heat pipe is an example of a closed system (within a system) that can remove heat from a hotter area. So, again, the "total system" must be determined in any of these problems.
3. Look at both steady-state and transient (start-up) conditions. Most systems will take some time to establish an equilibrium temperature. However, some systems have changing heat input, or changing outside ambient conditions, that must be taken into consideration. This text will consider only the steady-state condition as a general simplification.
4. Determine the ambient conditions that are outside the electronic enclosure. What are the temperature and altitude (pressure) highs and lows?
5. Determine the requirements (maximum temperatures) for both the individual components and the overall system.
6. Be prepared to include in the determination, the trade-offs for cost and reliability for different cooling solutions.
7. Determine the approach to solve the problem. Analytical programs exist to characterize the geometry and heat loading and provide answers with the assumptions as stated. "First-principle" calculations can also be made to get viable results. Another viable solution is to look at tables/graphs/examples from component manufactures that are provided in their literature or online. A testing plan must also be determined to verify the proposed solution.

As a general rule of thumb, typical heat dissipation per cubic inch of volume for an electronic enclosure (with the enclosure exterior cooled by free convection only) is generally:

- Open enclosure (holes allowing air in and out):   1 watt/in^3 free convection
- Closed enclosure (sealed – no holes):   0.25 watt/in^3 free convection

Note that a "typical" cell phone (3/4 watt in about 4 in^3 = 0.2 watt/in^3) is a good candidate for using free convection. A cell phone is a relatively "closed enclosure" with very limited openings for air in or out. A "typical" 14–15-inch laptop computer (60 watt in about 150 in^3 = 0.4 watt/in^3) should need something beyond free convection. Most laptops are cooled by a fan (forced convection) or possibly a heat pipe (which would be functioning closer to a heat sink in this example). Of course, laptop cases can't be so hot as to create an uncomfortable environment (hot, to the touch), and their components will be much more reliable with as much heat transfer as possible. Many devices use optimized processing performance through various power saving measures to achieve the ideal thermal situation – namely, an isothermal state over the product's outer enclosure.

Other typical heat dissipation rates per cubic inch of volume for electronic enclosures are:

- Forced convection (fan circulating air in sealed box): 1 watt/in^3
- Conduction only (sink attached to heat source, sink forming box exterior): 2 watt/in^3

Ultimately, portable devices must be cooled by combined natural convection/radiation heat transfer in what is called passive cooling. Passive cooling sets the upper bound on the system-level power consumption.

8.2   Conduction Mode of Heat Transfer

A sample problem is taken from Ref. [1], P. 39. It is typical to attempt to determine the surface temperature of a component (in this example, a transistor case) to make sure it is below the manufacturer's recommendation for maximum case temperature (see Fig. 8.1). Note that both radiation and convection effects are small (compared to conduction) in this example, and we'll see the order of magnitude for both radiation and convection in some further examples in this section. The input heat loading is specified as 25.6 Btu/hr. In this example, the change in temperature is computed across each "interface" starting with the known temperature of the chassis heat sink wall (55 °C). For each interface, the equation:

$$\Delta t = QL / KA \text{ is used where :}$$

Q is the heat load.
L is the thickness of the interface material.
K is the thermal conductivity of the interface material.
A is the cross-sectional area along the heat flow path.

The section being evaluated consists of:
Chassis heat sink wall
0.010 inch thick epoxy cement

$$\Delta t = \left(25.6\,\text{Btu} / \text{hr} \times 0.000833\ \text{ft}\right) / \left(0.167\,\text{Btu} / \text{hr ft}^\circ\text{F} \times 0.00677\ \text{ft}^2\right)$$

$$= 18.9^\circ\text{F}\left(10.5^\circ\text{C}\right)$$

Aluminum bracket

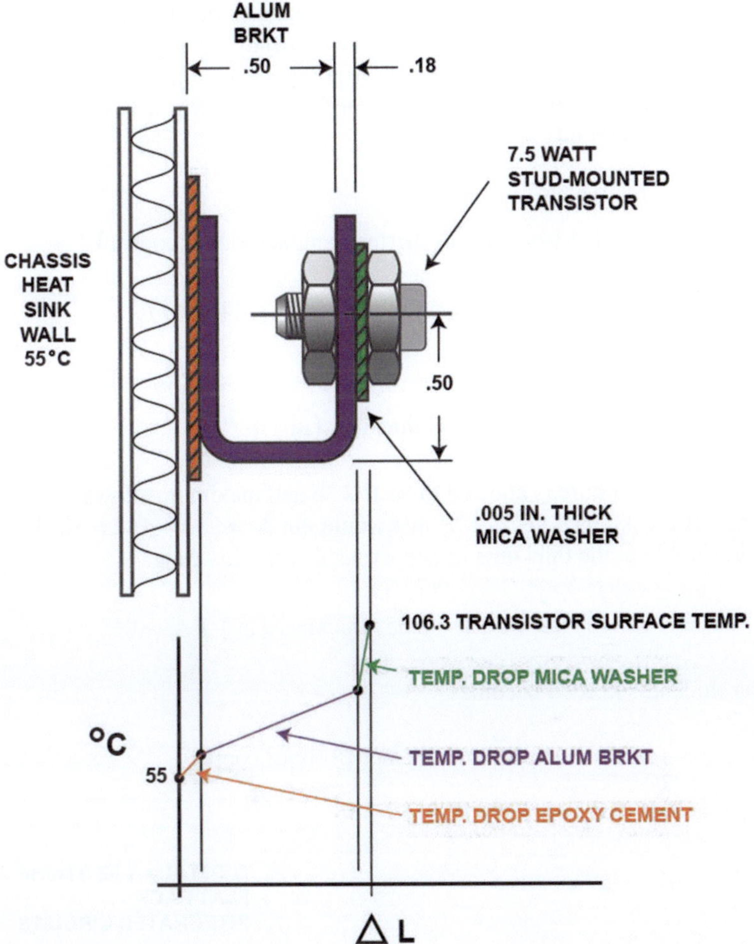

Fig. 8.1 Conduction example-Transistor

$$\Delta t = \left(25.6\,\text{Btu / hr} \times 0.125\,\text{ft}\right) / \left(83\,\text{Btu / hr ft}^\circ\text{F} \times 0.00097\,\text{ft}^2\right)$$

$$= 41^\circ\text{F}\left(22.8^\circ\text{C}\right)$$

0.005 inch thick mica washer (R, thermal resistance given from data book as 2.4 °C/ watt)

$$\Delta t = Q \times R = 7.5\,\text{watts} \times 2.4^\circ\text{C / watt} = 18^\circ\text{C}$$

So, the transistor surface temperature is the sum of the Δt in the heat flow path as:

$$55^\circ\text{C} + 10.5^\circ\text{C} + 22.8^\circ\text{C} + 18^\circ\text{C} = 106.3^\circ\text{C}$$

Another problem taken from Ref. [1], P. 44, calculates the thickness of 0.20-inch-wide copper needed in a PCB (printed circuit board) to dissipate a series of flat pack integrated circuits with a $Q = 1.02$ Btu/hr heat input (See Fig. 8.2).

Again, using $\Delta t = QL / 2KA$ (for max. temp rise in a strip with a uniformly distributed heat load (P. 44):

$$L = 0.25\,\text{ft}$$

$$K = 166\,\text{Btu} / \text{hr ft}^\circ \text{F} \left(\text{thermal conductivity of copper} \right)$$

$$A\left(\text{ft}^2\right) = 0.20\,\text{inch} \times \text{thickness}\left(\text{in.}\right) / 144\,\text{in}^2 / \text{ft}^2$$

$$\Delta t = \left(1.02 \times 0.25\right) \times 144 / \left(2 \times 166 \times 0.20 \times \text{thickness}\right)$$

$$= 0.55 / \text{thickness}\left(\text{in.}\right)^\circ \text{F}$$

If heat sink temperature is allowed to be 162 °F and maximum allowable case temperature (of the flat pack) is 212 °F, that would put $\Delta t = 212 - 162 = 50$ °F.

That would put the thickness of copper needed for this Δt at:

$$\Delta t = 50^\circ \text{F} = 0.55 / \text{thickness}\left(\text{in.}\right)^\circ \text{F}$$

Copper thickness needed becomes:

$$0.55 / 50 = 0.011\,\text{inch}\left(8\,\text{ounce copper}\right)$$

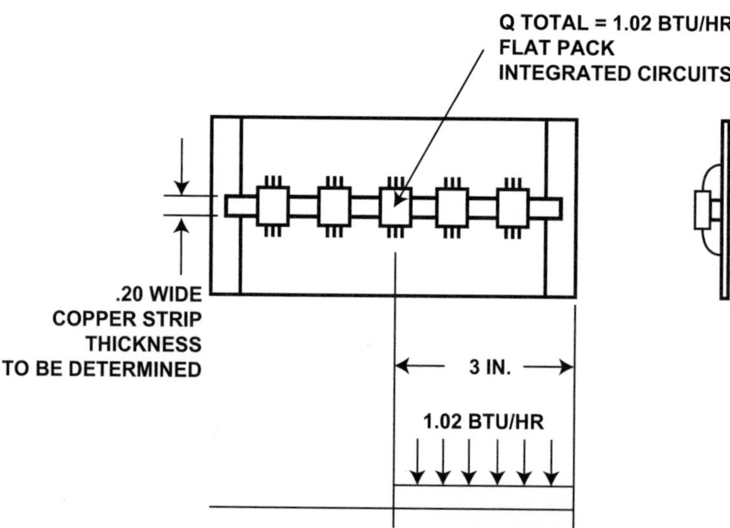

Fig. 8.2 Conduction example – Copper strip

Attaching a heat sink to a semiconductor package requires that two solid surfaces be brought together into intimate contact. Unfortunately, no matter how well prepared, solid surfaces are never really flat or smooth enough to permit complete, line-to-line contact. All surfaces have a certain roughness due to microscopic hills and valleys. Superimposed on this surface, roughness is a macroscopic non-planarity in the form of a concave, convex, or twisted shape. As two such surfaces are brought together, only the hills of the surfaces come into physical contact. The valleys are separated and form air-filled gaps. When two typical electronic component surfaces are brought together, less than one percent of the surfaces make physical contact. As much as 99% of the surfaces are separated by a layer of interstitial air. Some heat is conducted through the physical contact points, but much more has to transfer through the air gaps. Since air is a poor conductor of heat, it should be replaced by a more conductive material to increase the joint conductivity and thus improve heat flow across the thermal interface.

Several types of thermally conductive materials can be used to eliminate air gaps from a thermal interface, including greases, elastomers, and pressure-sensitive adhesive films. In the above problem (cooling the transistor case), this was the purpose of the addition of the 0.005-inch-thick mica washer.

8.3   Natural Convection Mode of Heat Transfer

A typical natural convection without radiation cooling application is where a heat sink is located in an enclosure, without any drafts, and the finned surfaces are "facing" other electronic equipment whose operating temperature is comparable to the heat sink temperature. The choice of heat sink contains a lot of variables, including ambient temperature, altitude, fin geometry, air velocity, density, viscosity, specific heat, thermal conductivity, and a host of dimensionless numbers (Reynolds, Prandtl, Grashof, Nusselt, etc.). Also, many of the variables do not relate to each other in a linear fashion.

Let's start the problem without any heat sink, first seeing if a heat sink is required (see Fig. 8.3).

We'll start with a TO-3 transistor component that has a resistance from its die (junction) to its case of 1.5 °C/W, a maximum junction temperature of 200 °C and a power dissipation of 150 watt (maximum). This will equate to:

$$Pd = \left(1 \,/\, \Theta ja\right)\left(Tjunc - Tamb\right)$$

If Tamb = 50°C, then Tjunc will then become:

$$150 = \left(1\,/\,1.5\right)\left(Tjunc - 50\right), \text{or } T\text{junc} = 275^{\circ}C$$

Thus, with the resulting Tjunc (without heat sink) higher than its 200 °C limitation, we will need to add some "resistance" into the thermal path that is bring the Θja to a higher value. This is done by adding a heat sink.

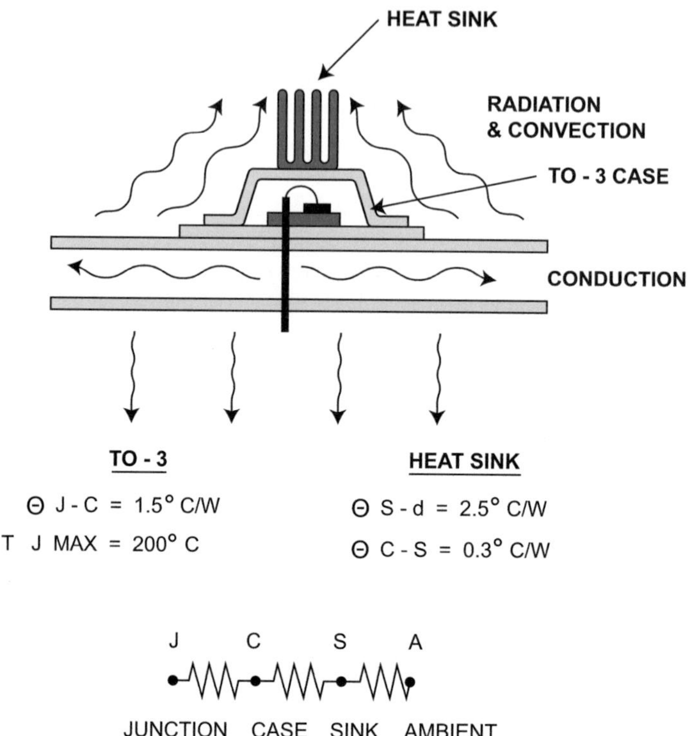

Fig. 8.3 Natural convection heat sink addition

Taking a look at a "network of resistors," we will add a heat sink which will add two resistances (the TO-3 case to heat sink resistance and the TO-3 to ambient resistance).

See Figs. 8.1, 8.2, and 8.3.

The power dissipation (Pd) depends on sum of thermal resistance from the component junction to ambient. Typically, ambient air temperature ($Tamb$), the maximum allowable junction temperature ($Tjunc$), and the thermal resistance for the component from junction to ambient (Θja) are related mathematically by:

$$Pd = (1 / \Theta ja)(Tjunc - Tamb)\,\text{and,}$$

$\Theta ja = \Theta jc + \Theta sa + \Theta cs$, that is, the resistance from junction to ambient is made up of the resistance of the junction to the TO-3 case, the TO-3 case to heat sink, and from the heat sink to the ambient.

Usually, both Θjc and $Tjunc$ are known from manufacturer's specifications.

Maximum ambient temperature, in this example, again, is 50 °C.

In our example of a TO-3 component, $\Theta jc = 1.5$ °C/W, and from the heat sink manufacturer (for that choice of heat sink), $\Theta cs = 2.5$ °C/W and $\Theta sa = 0.3$ °C/W,

that gives $\Theta ja = \Theta jc + \Theta cs + \Theta sa = 1.5 + 2.5 + 0.3 = 4.3$°C/W

If the Tjunc is specified as 200 °C, which would give the maximum power dissipation at:

$$Pd = (1/4.3)(200 - 50) = 34.9\,\text{W}.$$

It would be desirable to run at something cooler that the Tjunc (max. allowable), so, with a Pd of about 20 W, that would bring down Tjunc to about 136 °C, which would be a more reliable component.

8.4   Forced Convection Mode of Heat Transfer

In some cases, the heat loading is high enough to warrant forced convection as the main mode of cooling. This generally means the addition of one or more fans. Usually, forced convection is not considered for an environmentally sealed box (no openings). Let's go over some general considerations for the addition of a fan (air mover) to the system to cool the component(s):

- Flow (CFM) of air needed.
- General path of air and ducting required, air inlet and exhaust, and air filtration need.
- Maintenance and reliability of the fan.
- Acoustical considerations – if a small, high-speed axial flow fan is selected, the noise generated may be objectionable to personnel working in the area.
- Push or pull system (air being drawn into the system or exhausted out of the system).
- Pressure drop must be evaluated in order to minimize fan size, input power, and cost.
- Expected ambient environmental conditions such as altitude.

All of the above considerations make the fan choice a study of trade-offs and will likely need testing for the best solution. The usual design methodology of the following steps will result in success:

1. Description of problem and definition of success
2. Analytical computation (design) for first prototype proposal
3. Design review
4. Prototype testing and analysis
5. Repeat of above until problem is solved to specification

A "typical" analytical method proceeds as follows, with the following assumptions (for a typical electronic enclosure) problem:

Component hot spot temperature maximum = 212 °F
Maximum fan air inlet temperature (ambient conditions) = 131 °F
Maximum allowable air exit temperature (from the box) = 160 °F

Thus, allowable cooling air temperature rise = 160 − 131 = 29 °F
Power dissipation for (this problem's) fan = 25 watt
Power dissipation from (this problem's) electronic components = 140 watt (which, with 7 PCBs, is 20 watt (68.3 Btu/hr per PCB)
"Duct area," for this problem, given at 0.00625 ft^2

Phase 1 of problem is to find the temperature rise just due to the heat input from the fan and the electronics,

$$Q = 25 + 140 = 165\,\text{Watts} = 563\,\text{Btu} / \text{hr}$$

$$Q = m\,Cp\,\Delta Ta$$

where:

Q = power dissipation (heat)
m = mass flow rate = $V\rho$ (volume flow rate × ambient density)
Cp = specific heat of air = 0.24 Btu/lb °F
$\Delta Ta = Tout - Tin$, air temperature rise in box = 29 °F

Thus, the cooling air flow through the box must be:

$$m = Q / Cp\,\Delta T = 563\,\text{Btu} / \text{hr} / \left(0.24\,\text{Btu} / \text{lb}^\circ\text{F}\right) \times 29^\circ\,\text{F} \times 60\,\text{min/ hr}$$

$$= 1.35\,\text{lb} / \text{min}$$

Phase 2 of problem is to find the temperature rise across the convection film from the cooling air to the surface of the component. The convection relation is:

$$\Delta Tb = Q / hc\,A$$

$$Q = 68.3\,\text{Btu} / \text{hr}\left(\text{from above}\right)$$

$$A = \text{Area of PCB, given at}\,0.65\,\text{ft}^2$$

hc, the convection coefficient, is different for each forced convection case.
It is a function of:

1. The geometry of the air duct formed by the air space between the PCBs
2. Cross-sectional area of the above duct
3. Cooling air flow (determined above to be 1.35 lb/min). This generally gives the air flow needed (ft^3/min) that allows selection of fan. A fan can be selected to supply this flow rate, but that will be a flow rate where the fan impedance matches the box impedance at the system operating point. More on this concept at a later point in the discussion.
4. Properties of the fluid (air, in this case) at the relevant temperatures, including viscosity, specific heat, thermal conductivity
5. Whether laminar of turbulent flow results (determined by value of the Reynolds Number)

In our example, the resulting hc is:

$$hc = 6.22\,\text{Btu}\,/\,\text{hr}\,\text{ft}^{2°}\text{F}, \text{ and thus,}$$
$$\Delta Tb = 68.3\,/\,6.22 \times 0.65 = 16.9°\,\text{F}$$

The component surface temperature, $Ts = Tin + \Delta Ta = \Delta Tb = 131 + 29 + 16.9 = 1$
76.9 °F

Since the Ts is less than the maximum hot spot surface temperature (212 °F), this fan selection (and resultant flow) is sufficient.

Electronic boxes that are cooled with the use of fans must be carefully evaluated to make sure the fan will provide the proper cooling. If the fan is too small for the box, the electronic system may overheat and fail. If the fan is too big for the box, the cooling will be adequate, but the larger fan will be more expensive, heavier, and will draw more power.

Phase 3 of the problem is to determine that the static pressure being delivered by the fan (at particular flow rate) matches the flow impedance of the box (at *that* particular flow rate). Higher flow rates result in higher static pressure drop in the electronic box system.

Air flowing through the electronic box will encounter resistance as it enters different chambers and is forced to make many turns. This flow resistance is approximately proportional to the square of the flow rate in cubic feet per minute (cfm). When the static pressure drop of the air flow through a box is plotted against the cfm air flow rate, the result will be a parabolic curve. This "box impedance curve" goes from zero (zero flow and zero static pressure drop), to higher values of static pressure drop and flow rates. Meanwhile, the "fan impedance curve" goes from (high) value of static pressure drop, to a zero pressure drop at some high flow rate. Thus, these two curves cross at some common point of static pressure drop and flow rate. Where these two curves cross is the "system operating point", and this is how the system will operate, that is, there will be a particular flow rate and a particular static pressure drop.

An excellent analysis of determining whether a fan can deliver a (needed) flow rate at the fan's available static pressure loss is shown in Steinberg's text (see Ref. [1]).

8.5   Radiation Mode of Heat Transfer

A sample problem will be analyzed with and without radiation mode effects, as an introduction to the radiation mode.

Thermal radiation is the transfer of heat by electromagnetic radiation, primarily in the infrared wavelengths. The rate of heat transfer by radiation (from a surface) is mainly a function of:

- The emissivity of that surface
- The area of that surface

- The fourth power of the absolute temperature differential between the surface under question and the surface that is being radiated to (temperature is in "absolute" or degrees Rankine)

This usually takes the form of:

$$Q\text{rad} = S\, A\, \varepsilon\, \sigma\, \Delta T$$

S = "shielding factor," or "shape factor," ratio of energy intercepted by the cold body,
= 1.0 (maximum value) when radiating body is completely surrounded by body that heat is being generated to. See Ref. [2] for calculating values where the geometry gives values less than 1.0.

ε = emissivity of the surface. This is a de-rating factor for surfaces which are not black bodies ($\varepsilon = 1.0$). It should be noted that the term has little to do with the color in the optical sense; bodies of any optical color can have high emissivity and be referred to (thermally) as "black bodies." Also, be careful as to make a distinction between shortwave (solar) absorptance values and longwave emittance values. For example, white (oil-based paint) has a low solar absorptance value (0.25), while it has a high (0.90) longwave emittance. This distinction means that some surfaces will get hotter in sunlight than other surfaces. Also, material (and *finish*) matters quite a bit with emissivity values, for example, polished (or rough) aluminum sheet as an emissivity value near 0.05 while *anodized* (any color) aluminum as an emissivity value at 0.80.

σ = the Stefan-Boltzmann constant = 1.714×10^{-9} Btu/hr ft^2 R^4

For example, to examine the relative effect and to find the "heat lost", Qrad of a box, say with:

$$A = 6.3\,\text{ft}^2$$

$$S = 1.0\left(\text{box surrounded by ambient}\right)$$

$$\varepsilon = 0.8$$

$$\Delta T = T\text{surface} - T\text{ambient} = 20\,^{\circ}\text{C} = 68\,^{\circ}\text{F} = 459.7 + 68 = 527.7\,^{\circ}\text{R}$$

$$Q\text{rad} = S\, A\, \varepsilon\, \sigma\, \Delta T^4$$
$$= 1 \times 6.3 \times 0.8 \times 1.714 \times 10^{-9} \times \left(527.7\right)^4$$
$$= 670\,\text{Btu}\,/\,\text{hr} = 196\,\text{watt}$$

$$h\text{rad} = Q\text{rad}\,/\,A \times \Delta T = 670\,\text{Btu}\,/\,\text{hr}\,/\,6.3\,\text{ft}^2 \times 68\,^{\circ}\text{F}$$
$$= 1.6\,\text{Btu}\,\text{ft}^{2\,^{\circ}}\text{F}\,/\,\text{hr}$$

If one would analyze this same electronic box for the natural convection heat losses, say just the top surface, the same assumptions would lead to:

Qconvtop = hc Atop ΔT where:

Atop (of above radiation example) = 1.4 ft^2 (entire surface area of box was equal to 6.3 ft^2)

$$A\text{bott} = A\text{top}$$

$$A\text{sides} = 3.45\,\text{ft}^2$$

$$\Delta T = T\text{surface} - T\text{ambient} = 68\,^{\circ}\text{F}$$

$$hc = 0.27\left(\Delta T\,/\,L\right)^{\frac{1}{4}}\left(\text{from Steinberg Reference}\right),\text{where}$$

$$L = \text{"characteristic length"} = 2 \times 1.4\,\text{ft}^2\,/\,2.4\text{ft} = 1.17\,\text{ft, then}$$

$$hc = 0.27\left(68\,/\,1.17\right)^{\frac{1}{4}} = 0.27 \times 2.76 = 0.75\,\text{Btu}\,/\,\text{hr}\,\text{ft}^{2\circ}\text{F}$$

$$Q\text{convtop} = hc\,A\text{top}\,\Delta T = 0.75 \times 1.4 \times 68 = 71\,\text{Btu}\,/\,\text{hr}$$

$$Q\text{convbott} = hc\,/\,2\,A\text{bott}\,\Delta T = 35\,\text{Btu}\,/\,\text{hr}$$

For the bottom surface, it is roughly half of the top surface, so that is an additional 35 Btu/hr, while the (4) vertical sides (which totals to the entire six-sided box) have: equation below should be raised to the "1/4" power

$$hc = 0.29\left(\Delta T\,/\,0.7\right)^{\frac{1}{4}} = 3.6\,\text{Btu}\,/\,\text{hr}\,\text{ft}^{2\circ}\text{F}$$

$$Q\text{convsides} = hc\,A\text{sides}\,\Delta T = 3.6 \times 3.45 \times 68 = 845\,\text{Btu}\,/\,\text{hr}$$

So, for the entire six-sided box, natural convection is approx. $71 + 35 + 845 = 951$ Btu/hr.

Note that the heat lost due to radiation is approx. 670 Btu/hr. So, radiation didn't (and, doesn't usually) dominate the heat loss mechanism. If there were a steady breeze over the box (say, created by a fan), we would have the forced convection effects dominate the radiation effects even more.

8.6   Other Cooling Techniques

As power densities have gone up (more power needed in less volume), the need for "advanced" cooling techniques has risen.

Thermoelectric coolers offer the potential to reduce chip-operating temperatures or to allow higher module powers. Thermoelectric coolers offer the advantages of being compact, quiet, and free of moving parts, and their degree of cooling may be controlled by the current supplied. In recent years, there has been an increased interest to improve their performance through the development of new materials and thin film micro coolers. Cooling power densities above 100 watts / liter (1.6 watts/in^3) are now common.

Heat pipes are efficient and effective heat conductors. The thermal conductivity of a heat pipe is several hundred times greater than the thermal conductivity of the copper or the aluminum under the same cross-sectional area and thermal load conditions. Heat pipes transfer heat in a vacuum-tight vessel by:

- Working fluid (helium, nitrogen, liquid metals, water) is vaporized at the hot end, and a pressure gradient forces the vapor to flow to the cooler end where it is condensed at the cold end and returned to hot end by capillary effect through the wick surface.

Heat pipes channel the heat away, eliminating the hot spot. They are:

- Small sized
- Noiseless
- Flexible to fit in tight spaces
- Work in any orientation
- Maintenance free, no moving parts, high in reliability

and thus have been used in notebook computers and cell phones for years. They can remove about 20 watts of heat in at a ΔT of about 10 °C.

An ongoing good reference for advances in these cooling techniques is Ref. [3].

8.7 Thermal Testing

Detailed thermal analysis is required as more mechanical and electrical details become available during the design cycle. The mechanical design engineer first determines if natural convection/radiation is enough to cool the system. Hand calculations, prototype mockups, and very simple CFD (computational fluid dynamics) models can all be used. Detailed thermal analysis is required as more data becomes available. Uncertainties on the temperature estimates can be at +/−10% at the beginning stages of the design as device power dissipation and component thermal properties can be sources of error. To help reduce these uncertainties, prototype temperature and airflow testing are done as soon as hardware is available. A portable thermal camera, thermocouple data logger, and airflow test equipment should be available to allow testing. Since device power estimates can be one of the greatest sources of errors in the thermal analysis, the power estimates can be verified at early prototype stages by comparing infrared images or thermocouple measurements to predicted values.

Thermistors can be mounted as components on PCBAs to monitor temperature.

Web-hosted applications for specific thermal analysis tasks such as:

- Heat sink design
- Component thermal analysis

are available from sources such as R-Theta or Flomerics.

A manual on the use of thermocouples is sourced as Ref. [4].

References

1. Cooling techniques for electronic equipment, by Dave S. Steinberg, John Wiley & Sons Publications.
2. Principles of heat transfer, by Frank Kreith, Intext Educational Publishers.
3. Electronics cooling., www.electronics-cooling.com, by Andara Print.
4. Manual on the use of thermocouples in temperature measurement, ASTM.

Chapter 9
EMC

After starting with some definitions, we will focus on this major consideration of the overall enclosure design. The design of *all* electronic enclosures will need to consider two things:

1. Is the electronic product being designed going to operate properly while being affected ("bombarded") by self-generated electromagnetic waves?
2. Is the electronic product generating electromagnetic waves that will negatively affect other products?

All electronic products, again these are what the EPE Designer is designing, are *susceptible* to interference, and they are *generating* interference themselves. There are regulations for the frequency specific levels that all electronic products must meet for this generated interference. We'll discuss what these levels are.

We will go over various techniques that are used by the EPE Designer to shield an enclosure and to go over the trade-offs in making some final choices between techniques.

We'll close with a discussion of the existing worldwide EMC standards.

9.1 EMC Issues

So we can begin the general topic, let's start with defining some terminology: (From Wikipedia)

Electromagnetic interference (EMI), also called **radio-frequency interference (RFI)** when in the radio-frequency spectrum, is a disturbance generated by an external source that affects an electrical circuit by electromagnetic induction, electrostatic coupling, or conduction. The disturbance may degrade the performance of the circuit or even stop it from functioning. In the case of a data path, these effects can range from an increase in error rate to a total loss of the data. Both man-made

© Springer International Publishing AG, part of Springer Nature 2019

T. Serksnis, *Designing Electronic Product Enclosures*,

https://doi.org/10.1007/978-3-319-69395-8_9

and natural sources generate changing electrical currents and voltages that can cause EMI: automobile ignition systems, cell phones, thunder storms, the Sun, and the Northern Lights. EMI frequently affects AM radios. It can also affect cell phones, FM radios, and televisions.

The more general term, Electromagnetic Compatibility (EMC), which is preferred by the electronics industry (IEEE), shall be used in this chapter going forward for the general problem for designing products that need shielding against EMI/RFI.

EMC is thus defined as the ability of a device, equipment, or system to function satisfactorily in its electromagnetic environment without introducing intolerable electromagnetic disturbances to anything *in* that environment.

Designers of products that contain electronic circuits, or basically need a power supply, have two problems related to EMC:

1. How to shield internal components (that are susceptible to EMI) from that interference. The interference could be generated from a completely *external* device (from another product) or could be generated from another device *within* the product (fratricide). This is generally labeled here as the "susceptibility problem."
2. How to shield the product so that it doesn't interfere with *other* products and comply with rules and regulations that prohibit that interference. Some major standards (FCC and VDE) will be discussed more fully in Sect. 9.3.

The two problems above can be thought of as separate but related. The usual tact is to get the product functioning (not interfering unacceptably with itself), and then submit the product to testing to pass the industry standard for interference with other products. Most companies would have an internal testing program where the prototype would pass "internal (to the company) testing" before being sent out to a contracted "external" testing facility that could certify that the product did indeed pass the industry standard and, as an end result, be certified as passing the testing. Some degree of "self-certification" is possible, and FCC/VDE has specific rules that apply to these cases. As certification is required showing the passing of government regulations, it becomes a time-critical element to:

1. Produce a representative prototype(s) for submission to the agency approval office for testing.
2. Allow enough time for the scheduling at the agency approval office for testing.
3. Provide the test time needed for actual approval, and notification that product shipment to customers is permitted. This would permit the correct labeling (agency approval logo, indicating test approval) to be placed on the product.

Just as there is no decision made in the design of electronic products that doesn't involve "structural considerations," or "thermal considerations," EMC issues will also *dominate* various design choices. Just like "structural" or "thermal" issues, the "EMC issue" will affect *cost* trade-offs and therefore be an integral factor in overall design decisions. As with "structural" or "thermal" issues, the design that includes "EMC" will proceed as:

1. Determine overall design goals (time/budget/specification).
2. Design using analytics, experience, prototypes, and review procedures.
3. Submit product samples to a testing program and pass test specifications.
4. Determine "lessons learned" from project as basis to start next project.

Various references on the subject help immensely with information on the subject of EMC. References [1, 2] are "must reading" for mechanical design engineers. Other valuable sources are listed and are mostly gathered from shielding material manufacturers.

It is clear that the problem of understanding and optimizing the solution to any EMC issue will take the *combined* efforts of Digital, RF, and Mechanical Engineering. Reference [2] gives "The Pragmatic Approach: Shielding Evaluation of a Prototype Cabinet." This states: When the exactitude of the design analysis is critical, because the cost impact in "$ per dB" is high, analytical calculations may not be the best overall approach. A more straightforward approach is to do a test evaluation of the planned enclosure. This would require:

1. A prototype of the designed enclosure with all of the holes, seams, and openings as representative as possible
2. A radiating source, preferably battery operated, to avoid power cord radiation which could be mistaken for shield leakage
3. An EMI receiver or spectrum analyzer and its antenna

The radiating source is located inside of the prototype enclosure, and a plot of the received signal (at say 1–3 meter distance) is made. Comparison is made between readings with or without the enclosure, and the difference in dB is the shielding attenuation of the prototype enclosure.

Reference [1] gives an overview of the *Shielding Design Methodology and Procedures* with a very well thought-out process of:

1. Determining *shielding effectiveness* (SE) requirements. Typical units for SE end up in dB (decibels). These can be determined from several sources, and representative radiators can be identified that simulate the environment across a frequency spectrum. Typical ambient levels from licensed transmitters would look like (e.g., the 100 watt paging systems):

Electric field strength (volts/meter)	Transmitter-to-victim distance (meters)
3.00	30.0
0.29	300
0.09	1000

If shields were perfect, both the electric and magnetic field energy escaping the enclosure would be zero. The shielding effectiveness would be at its highest level. Shields perform on 2 principles:

- Absorption, which increases with: thickness/conductivity/permeability/frequency
- Reflection, which increases with: surface conductivity/wave impedance

To evaluate reflection, it is necessary to know if the shield is in near- or far-field conditions. The near-field conditions are the most critical. For pure electric fields, since their wave impedance is high, it is relatively easy to get good reflection properties because the field-to-shield mismatch is large. For near magnetic fields, the wave impedance is low and it is more difficult to get good reflection.

At low frequencies, what counts is the nature of the metal which is used: thickness/conductivity/permeability. At higher frequencies, where any metal would provide hundreds of dB of shielding, they are never seen because seams and discontinuities completely "spoil" the metal barrier.

2. Determining victim susceptibility. This would be taking those values from step 1 above, adding the "field-to-cable coupling, subtracting the "sensitivity," and adding in "margin of safety." Both steps 1 and 2 result in getting the overall SE for the product (in dB).
3. Determining the "Box Leakages." Box Leakages are all of those areas of the product where EMI is expected to "leave the product." In a metallic enclosure, every opening (non-conductive gap) must be examined. In a nonmetallic enclosure (i.e., plastic), that enclosure must be made metallic (conductive) by some means, and again, every opening must be examined. The main design objective for compliance to EMC requirements will be to *minimize* (the size and number of) *these openings*. Typical openings are:

- Displays (LED/CRT)
- Keypads or Membrane Switches
- Status Indicators
- Cooling Apertures
- Box Assembly Seams
- Cable or Connector Openings

Electromagnetic energy hits an enclosure, and some amount of that energy is absorbed, while some is reflected back internally. The amount of energy absorbed or reflected depends on the frequency of the energy. Absorption loss plays a larger role at the higher frequencies.

$$\text{SEbox}\left(\text{dB}\right) = \text{Reflection}\left(\text{dB}\right) + \text{Absorption}\left(\text{dB}\right)$$

Shielding effectiveness is defined as the ratio of the impinging energy to the residual energy (the energy that gets through),

$$\text{For E fields,} \text{SEbox} = 20\log\left(\text{Ein / Eout}\right)\left(\text{dB}\right)$$

$$\text{For H fields,} \text{SEbox} = 20\log\left(\text{Hin / Hout}\right)\left(\text{dB}\right).$$

If shields were perfect, Eout and Hout would be zero. In practice, a shield performs on two principles:

- Absorption increases with thickness/conductivity/permeability/frequency.
- Reflection increases with surface conductivity/wave impedance.

At the end of the above analysis, all of the box leakages are totaled (in dB). With the:

1. Maximum dB levels (at specified frequencies) permitted for certification become the SE Objectives.
2. An "ideal box" (no leakages) performance is determined.
3. Individual "aperture leaks" are determined.
4. All of the individual "aperture leaks" are combined (in accordance combining equations for multiple aperture leaks).
5. The combined "aperture leaks" are subtracted from the "ideal box" performance (step 2 above, minus step 4 above). Let's call this result: SEactualbox.

If SEactualbox exceeds the SE Objectives, then an iterative process is begun to reduce the individual "aperture leaks" (from step 3 above) until:

$$SE\,Objectives > SEactualbox$$

9.2 Shielding the Enclosure

The text *Interference Control in Computers and Microprocessor-Based Equipment* (Ref. [2]) has chapters that cover:

- PCB Design and Layout – where (mainly) RF and Digital Designers will design the PCBAs to reduce or eliminate noise and susceptibility.
- Power Supply Design and Layout – where (mainly) Power Supply Designers will design the power supplies to reduce or eliminate noise and susceptibility.
- Internal Wiring – where the above Designers and Mechanical Designers will work to design interconnects between PCBAs and to the enclosure which will not pick-up or induce EMI internally (self-jamming) or externally (violate FCC/CISPR/VDE/Military regulations).
- Enclosure Design – where Mechanical Designers will design a *continuously closed conductive envelope* in order to prevent outside fields from penetrating the equipment and to prevent internally generated noises from escaping the enclosure.

As stated above in Sect. 9.1, the product will achieve EMC only when an overall design that integrates both electrical and mechanical disciplines is accomplished.

The following are discussions on techniques that have shown to be effective in designing the "continuously conductive envelope:"

It will be assumed that the "enclosure" is the *first* enclosure surrounding the active circuits (PCBAs). Enclosures *outside* of this *first* enclosure may (or may not) be conductive as EMC is assumed achieved by the *first* enclosure. Thus, a product like a cell phone can have a thin metal shield that surrounds the circuitry, while the external case may be plastic. The plastic case adds nothing to EMC, so, again, we would only be designing for EMC with this *first* enclosure (for the purposes of these examples). It is possible that this *first* enclosure doesn't *fully* achieve the EMC target, and is further *aided* by a *second* enclosure to achieve the EMC target.

The enclosure should be a continuous metal surface where PCBA grounds are attached at specific locations. This continuous metal surface can be either "actual" metal (like aluminum, magnesium, or steel) or it can be a conductive coating that has been put onto a non-conductive material (like plastic). Note that, when considering "actual" metal as an enclosure, the designer must take into account that the surface finish on the metal must be *conductive*. Several common surface finishes for metal are *not* conductive (enough) to serve as a "continuous metal surface." Any finish chosen to be conductive, must remain so during the environment (corrosion) and wear (say, due to vibration) that the product will undergo during its expected life.

Another way of providing this "continuous metal surface" is to use conductive filler interspersed in a non-conductive material (such as plastic). This essentially looks like metal to the EMI energy. Care must be taken to adequately ground past a non-conductive "skin" on this type of material.

Enclosures are "nominally" 5-sided boxes, with the *remaining side* (let's refer to the *remaining side* as the "access panel" or just the "panel") providing access for assembly and service. Thus, a "seam" is introduced into the design that must be designed so that the panel contacts the box "continuously." Some points to be made about the design options are here:

1. It is possible to introduce into the design a compliant conductive gasket which will set between the box and panel – this would be a "continuous" box if the gasket indeed contacted the surface of both the box and panel with enough pressure. Assuming that a bolt pattern is attaching the box and panel, the pressure is 100% at a bolt position, but something less than 100% between the bolts (due to the gasket). (More on this design situation further on in this chapter.)
2. If a compliant conductive gasket is not in the design, then it is only at the bolt positions that there will be (guaranteed) contact. In any area between bolts, as the surfaces of both the box and panel are not exactly flat, we can expect some degree of a "gap" (that allows EMI to escape the box). For wavelengths that are less than two times the longest aperture dimension (the length of the gap or opening), the electromagnetic energy will pass freely through the opening without being attenuated. For wavelengths equal to twice the opening, the shielding is *zero*. Reference [3] shows a graph of the relationship between opening size vs. frequency for both $\lambda/20$ (max. opening for commercial products) and $\lambda/50$ (max. opening for military products). So, openings larger than $\lambda/20$ would be avoided for commercial products. For frequencies at about 10 GHz, this opening size

(commercial products) is 0.1 cm (0.04 in.), while at a 100 MHz frequency, the opening size is about 7 cm (2.8 in.). Threat frequencies are typically greater than 100 MHz.

Reference [4] shows a graph that indicates maximum opening size is about 1cm (0.39 in.) to adequately shield at 100 MHz.

In cases were "overlap" occurs, that is, the panel and box overlap (do not touch), the two surfaces of the seam form a capacitor. (See Fig. 9.1) Since capacitance is a function of area, seam overlap should be made as large as practical to provide sufficient capacitive coupling for the seam to function as an electrical short at high frequencies. As a good rule to follow, the minimum seam overlap to spacing between surfaces ratio should be 5 to 1.

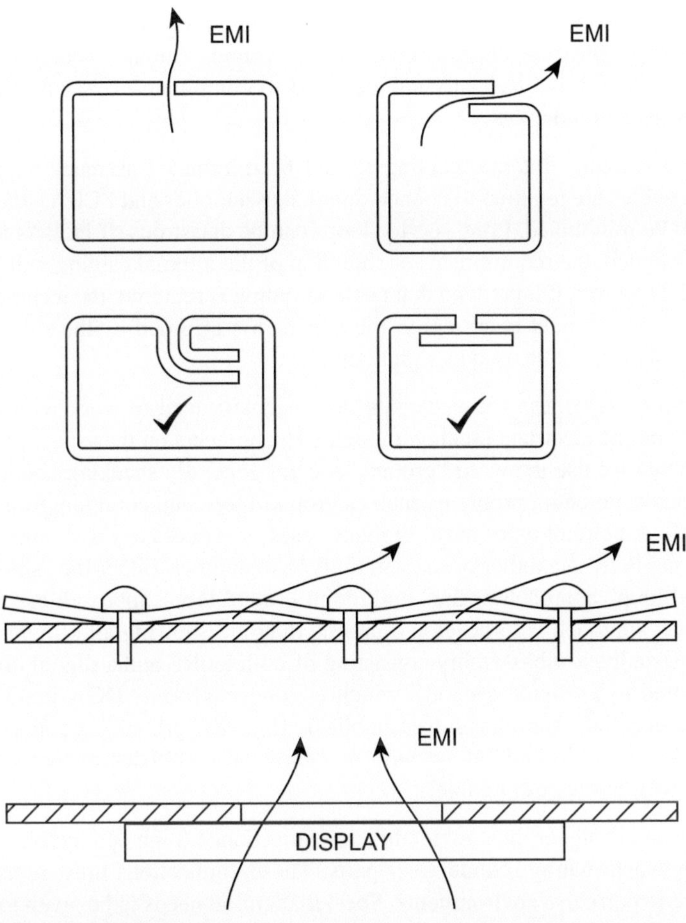

Fig. 9.1 Enclosure seam leakage

3. Most shielding material manufacture's guidelines (see Ref. [5]) have information about calculating bolt spacing so that EMI from the box is minimized. Trade-offs occur for assembly, cost, and shielding effectiveness. Box/panel deflection between fasteners is a complex problem involving the geometry, material of both the box and panel, and the asymmetrical application of forces in two directions.

4. Windows: Often, large-area openings in the "continuous metal surface" are required for viewing displays, status lamps, or other large non-conductive feature. Several options are available to make this area conductive and grounded back to the (metallic) case:

Laminating a conductive screen between optically clear plastic or glass sheets

Casting a mesh within a plastic sheet

Applying an optically clear conductive layer to a transparent substrate

5. Cable Penetrations: To maintain shielding integrity of the enclosure at cable penetrations, shielded cables must be used. On the exiting cable, a shield that surrounds the cable is attached back to "chassis ground." Another technique is to use a filter (in-line) that basically reduces EMI to compliant levels with the use of capacitors and inductors.

6. Internal Cabling: Internal cabling should be minimized as much as possible. When cables are required to connect internal assemblies and PCBAs, the lengths should be minimized. Long service loops can be disastrous. If PCBAs are properly designed, the requirement for shielding of the internal cabling will be minimized. However, if it is found that cable shielding is required, the technique used to ground the shield is critical to the attenuation afforded by the shield. Cable shields should not be used as signal returns.

7. Grounding Schemes: The enclosure designer will need to work with both the safety and the electrical (analog, digital, RF) engineers on the aspect of grounding. When we use the word "ground," we are generally speaking about a reference point. Potential problems such as ground loops and common ground paths for different circuit types exist. In many cases, it is necessary to connect returns from one PCB to another or one circuit type to another. Either the "single-point ground loop" approach or the "multipoint ground loop" approach can be used. The best approach is to develop a ground diagram showing all ground connections. Usually, a subassembly consisting of both an RF and a digital circuit are connected to a "signal ground," which is connected to a "DC ground" (along with connections for other subassemblies). This "DC ground" is connected to a "chassis safety ground" along with an "earth ground" back at the AC (green safety wire in the power cable).

The enclosure designer must maintain a low impedance (high conductive) connection between mating (conductive) parts. These connections must remain conductive in corrosive environments. Special attention needs to be given to the use of dissimilar metals to preclude the effects of galvanic action. The goal is to maintain, as close as practical, a single potential "safety ground" system. This is good practice for EMC as well since the "safety ground point" will also serve as the primary point of reference for all other ground connections.

9.3 EMC Standards

The reader is cautioned to get a copy of the latest standards mentioned in this section. The issue levels referred to here are meant to be a starting point for the reader.

9.3.1 Military Standards

In solving an EMC problem, both the emitter noise level and susceptor's noise threshold must be considered. If the susceptor's lowest signal threshold level can be made greater by at least two times the highest emitter (noise) level (for a 6 dB safety margin) then the emitter and the susceptor are considered to be compatible with each other. One way of achieving this type of compatibility is to define not only the maximum allowable emissions level but also the minimum allowable susceptibility threshold level. Systems that comply with both of these requirements will be compatible with all other systems which also meet these requirements.

The most commonly used military standard for both emissions and susceptibility is MIL-STD-461. There are more than one emission and susceptibility level defined in MIL-STD-461 since the requirement applications vary (e.g., aircraft vs. tracked vehicles vs. submarines).

9.3.2 Commercial Standards

Unlike military standards, commercial standards imposed by law deal only with emissions. The emission standards most applicable to commercial electronic equipment are Federal Communications Commission, FCC Part 15, Subpart J. (Docket 20780), which deals primarily with electronic data processing equipment, and the German Standard VDE 0871/6.78. The German Standard is in close agreement with the recommendations of the international committee, CISPR, and hence can be applied to most equipment to be sold in the European Economic Community.

Taking a look at my Hewlett-Packard (HP) Slate 21 Pro (All-in-One Computer) User Manual, it has the following regulatory notices, many of which are EMC regulatory in nature:

FCC Category B from the U.S.A.
Notices from Brazil, Canada, European Union, Germany, Australia, New Zealand, Japan, South Korea, Mexico, and Taiwan NCC
"CCC S and E" certification label from China

It takes a lot of experience to comply with the EMC standards in all of the world-wide markets. Many of the individual nations do not differ greatly in the levels that are to be met, but there are distinct differences that must be tested for.

Both the FCC and VDE standards have emission limits for Category "A" and "B" equipment. The corresponding categories, however, are not classified in the same way. In Germany (VDE), equipment that complies with VDE "B" limits does not require a verification test if the emission limits of Category "A" limits can be applied. However, qualification to Category "A" requires that a verification test be performed.

For the FCC, Category "A" refers to digital equipment that is to be used in a commercial, industrial, or business environment. Category "B" refers to electronic data processing equipment marketed for use in a home or residential environment. At present, only Category "B" equipment, consisting of electronic games of all types, personal computers (excluding handheld calculators and digital watches), and any device intended to be connected to a TV receiver or TV interface device, requires certification by the FCC. All other equipment is certified by the manufacturer.

As an example of the regulation limits, graphs show the attenuation (dB) as a function of frequency (MHz).

For FCC Category B (radiated):

Frequency (MHz)	Attenuation (dB)
40–88	40
88–216	43.5
216–1000	46

For FCC Category B (conducted):

Frequency (MHz)	Attenuation (dB)
0.45–3000	48

References

1. White D, Shielding Design – Methodology and Procedures. Interference Control Technologies, Gainesville
2. Mardiguian M, Interference Control in Computers and Microprocessor-based Equipment. Don White Consultants, Inc., Gainesville
3. EMC Compatibility Design Guide. Tecknit
4. Engineering Design and Shielding Product Selection Guide. Instrument Specialties
5. EMI Shielding Engineering Handbook. Chomerics

Chapter 10
Safety by Design

I decided to title this chapter, "Safety *by Design*" instead of just "Safety." I'm hoping this emphasizes the idea that safety must be *designed into* the product enclosure and, indeed, the entire product. Everyone who comes in contact with the enclosure, from assembler to tester, to customer, and to the service technician, must get the benefit of reasonable safety. A warning must be clear and emphatic.

We'll start the chapter with the requirement of formal safety reviews needed during the product development process. We'll review items that usually are reviewed, and these will be addressed by the EPE Designer. We close with a section covering the approval processes and safety standards that exist on a worldwide basis.

10.1   Safety Review

Safety in product design is of paramount importance. As in the physician's popular phrase "First, do no harm" (mistakenly referenced as part of the Hippocratic Oath), a product designer should think first about safety. In the life cycle of product development, safety issues occur several times. For example:

1. Testing that involves the product
2. Fabrication of individual parts and the assembly of those parts
3. The product itself as it functions (or malfunctions) in the hands of the consumer

In all cases their needs to be, as a part of the standard design process, periodic reviews to insure that the product itself and all parts of the product development do not cause injury or property damage. Each project must have a person who is responsible for a safety review. Some industries with obvious emphasis on safety concerns are health, military, space, transportation, food, public utilities, firearms, and lasers. Safety is, by itself, worthy of utmost attention, but the litigation from

© Springer International Publishing AG, part of Springer Nature 2019   181
T. Serksnis, *Designing Electronic Product Enclosures*,
https://doi.org/10.1007/978-3-319-69395-8_10

accidents can also cripple a corporation. The safety reviews should follow the guidelines as previously stated for "Design Reviews," Section 1.5 of this text.

In the design of electronic enclosures, these systems use a power supply, and that power supply (AC, DC, AC to DC conversion) needs scrutiny for electrical safety. Also, batteries themselves have additional transportation and disposal issues that must be carefully analyzed for safety concerns.

Any chemical used in the product must be identified as potentially harmful to people or the environment. This would include adhesives, solvents, and coatings used to make individual parts, or in the assembly of those parts. Materials that are considered a risk have been identified by Material Safety Data Sheets (MSDS) and have national and international requirements.

Per Wikipedia, a safety data sheet (SDS), material safety data sheet (MSDS), or product safety data sheet (PSDS) is an important component of product stewardship and occupational safety and health. It is intended to provide workers and emergency personnel with procedures for handling or working with that substance in a safe manner and includes information such as physical data (melting point, boiling point, flash point, etc.), toxicity, health effects, aid, reactivity, storage, disposal, protective equipment, and spill-handling procedures. SDS formats can vary from source to source within a country depending on national requirements.

RoHS, also known as lead-free, stands for Restriction of Hazardous Substances. RoHS, could also be referred to as Directive 2002/95/EC, originated in the European Union and restricts the use of six hazardous materials found in electrical and electronic products. All applicable products in the EU market after July 1, 2006, must pass RoHS compliance. RoHS impacts the entire electronics industry and many electrical products as well. Materials under restriction include lead, mercury, and cadmium. These materials are monitored and restricted due to their health and environmental detriments.

Every manufacturing step needs to be examined for potential safety issues. For example, if ultraviolet light is being used (say, to cure an adhesive), any harmful effects of UV needs to be understood and protected against.

All manufacturing processes (done in-house or done by a supplier) need to be examined for safety issues. If a part is to be manufactured by a supplier, the supplier should have safety procedures published and available. An example of this is shown in [1]. Anytime the fabrication process produces a sharp edge, as with burrs on sheet metal, that edge needs to be removed so that it is not a hazard to both the assembly personnel and the customer. Square edges should be chamfered or beveled so that edge points do not occur.

10.2 Mechanical Issues

For particular issues relating to enclosure design, see detail in UL 478, referenced in Section 10.3.

10.2.1 Temperature of Enclosure Surfaces

This includes both the external (customer accessible during operation) surface and operating devices such as handles. Areas that might get very hot, such as heat sinks, should be guarded from being customer accessible and have warning information if they would remain hot even if power is off or during servicing.

10.2.2 Physical Stability

A product shall not tip over when tilted 10° from its intended, upright position, while all doors, covers, gates, etc. are in place and closed. Also, for a "free-standing" unit, the product shall not tip over when a force of 180 pounds is applied at the point of maximum moment.

10.2.3 Mechanical Strength of Enclosures

Enclosures should not distort to cause safety issues when reasonable forces are applied. Reasonable impacts should not induce breakage that would cause safety issues.

10.2.4 Cable Strain Reliefs

All cables (internal and external) should be able to withstand reasonable pull and rotational twists that will occur during the life of the product.

10.2.5 Flammability Tests

Enclosure material shall not release flaming drops or flaming or glowing particles that ignite surgical cotton about 12 inches below the test sample and shall not continue to flame for more than 1 min after the application of a test flame. Designers should specify plastic that passes UL 94.

UL 94, the Standard for Safety of Flammability of Plastic Materials for Parts in Devices and Appliances testing, is a plastics flammability standard released by Underwriters Laboratories of the USA. The standard classifies plastics according to how they burn in various orientations and thicknesses, from lowest (least flame-retardant) to highest (most flame-retardant).

10.2.6 Labeling

The enclosure should have adequate labeling to warn the assembly, test, service, and end users of any unsafe condition. Agency approval labeling should be per that agency's specifications.

There are specific standards on labeling material and wording.

10.2.7 Batteries

All batteries should be installed and operated per manufacturer specification. Disposal, shipping, and labeling of batteries are to be done to encourage recycling, to follow the laws regarding recycling, and for the protection of the environment.

"Manufacturer specification" should have the original specifications from the OEM battery supplier and should include information given to the customer from the equipment (that the battery will operate in) manufacturer.

10.2.8 Grounding

Ground paths that depend on grounding integrity for safety should be designed to have these paths in contact thru the life of the product. These ground paths can be subject to galvanic corrosion and transportation shock and vibration.

10.2.9 Accessibility

Dangerous items in the enclosure should be interlocked, protected against, secured, or free of such items as:

- Harmful electrical energy (voltage, currents)
- Sharp edges
- Dangerous moving parts
- Temperatures greater than 60 °C
- Toxic chemicals
- Dangerously bright light sources
- Areas where fingers could get trapped or injured
- Harmful acoustic pressure levels

from both the consumer and service personnel.

10.2.10   Liquid Spills

Electrically live parts should be protected against the possible spillage of liquids onto the enclosure.

10.2.11   Conductive Coating Adhesion

Conductive coatings should be stable over the life of the product. These coatings should be able to pass UL testing for adhesion.

10.3   Approval Process and Safety Standards

The reader is cautioned to get a copy of the latest standards mentioned in this section. Standards are always in a constant state of change based on new technologies and new information. The particulars mentioned here are meant to be a starting point for the reader.

It appears that worldwide, the safety standards have generally standardized on some variation of "60950". Various standards are listed:

- Underwriter's Laboratory U.L. 478, now revised to UL 60950-1
- Canadian Standards Association C22.2
- International Electromechanical Commission (IEC) as IEC 60950-1
- European Union as EN 60950-1
- German Standards VDE 0804-100 / DIN EN 41003
- China Compulsory Certification (CCC) latest standards

It's clear that safety issues need to be considered both for their intrinsic need and for compliance to regulatory standards in the markets that the products will be sold into.

The need for "production units" for submittal to various agencies for agency approval will usually end up being in the critical path of product development. This occurs because the length of time needed for approval (submission of samples to agency approval) will be quite large (on the order of months).

Approval process proceeds as:

- Submission (Request for Quote, RFQ) is initiated either at the Approval Body or Contractor for services.
- Submission documentation and samples are created.
- Testing is conducted and (if passing).
- Approval is granted for customer shipment.

Reference

1. Guide to safety in the metal fabrication industry, WorkSafe, Victoria, Australia

Chapter 11
Shipping and Packaging

This chapter will concern itself with the design of the packaging that the product, as designed in previous chapters, will ship to the customer in. I'm adding this chapter as it seems to be a part of the total design process for the product, and in many cases comes under the purview of the EPE Designer, although an entire (separate) department may fulfill this effort in larger firms. In any case, the EPE Designer needs to be generally aware that their product must be shipped to the customer so that the highest customer satisfaction is attained, and this starts with the original unpackaging of the product (aka "the out-of-the-box experience").

A good current example of this is the packaging for a 50-inch flat-screen LED television from Visio Corporation that was purchased in 2015. The package that the television came in was a box with dimensions of

5 inch wide × 29 inch high × 47 inch wide (see Fig. 11.1)

Inside this box was an LED TV that needed to be assembled to its stand, "clamshell" dense-foam pieces (upper and lower) that protected the TV, a remote control, and a "quick-start" printed guide. The TV itself was surrounded by a sheet of thin "polyfoam" that protected it from scratches. Once an antenna (not included) was attached, and the TV was powered up via AC cord, it got signal and I watching my favorite channel. The TV came with internal Wi-Fi and attached to my in-home intranet. Thus, the "out-of-the-box experience" was excellent in that:

1. The product arrived in functional condition.
2. The product arrived without any cosmetic flaws.
3. The product was "up and running" with very minimal information needed by the customer (which was clearly stated in the "quick-start" guide). If customers need to call or contact the supplier with questions on "startup," the additional cost needed for this would be (eventually) added to the cost of the product. If no customer contact is required, overall costs are reduced, and the overall customer experience is likely improved.

© Springer International Publishing AG, part of Springer Nature 2019
T. Serksnis, *Designing Electronic Product Enclosures*,
https://doi.org/10.1007/978-3-319-69395-8_11

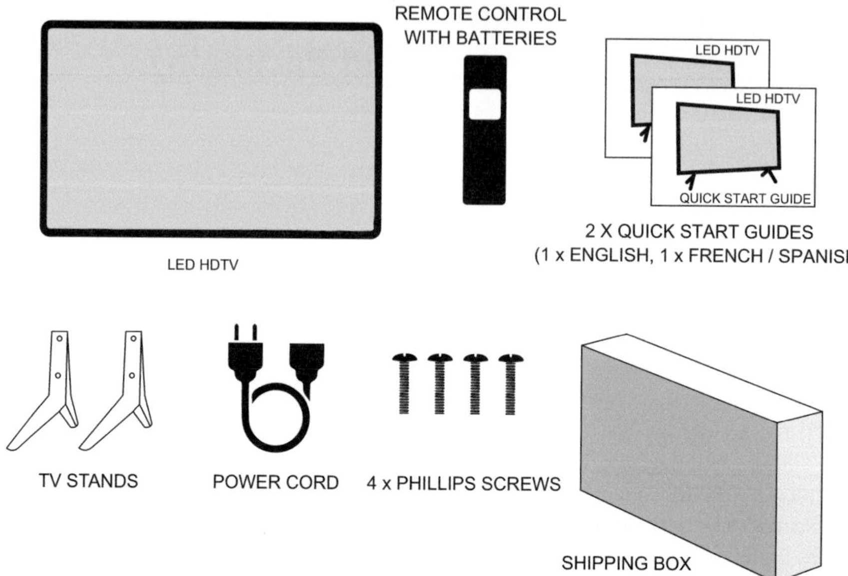

Fig. 11.1 In the box

4. Some very easy assembly was required (with hardware supplied, no special tools required for assembly, and extremely simple assembly instructions). This was the assembly of the TV to its mounting bracket (4 bolts total). The choice of how much to ask the customer to assemble is quite a "design problem" in itself. It clearly depends on tradeoffs between:

- Ease of assembly and customer expectations for assembly required.
- Overall outside box dimensions needed for expedient shipping of product. Sometimes, if the product is completely assembled, it can be more fragile or take up more volume, which ends up increasing the cost of shipping the product. Getting the overall dimensions of the shipping box reduced may add assembly for the customer but may reduce shipping costs.

Also note that the TV itself has dimensions of 2.5 × 26 × 44. This is approximately 42% of the total shipping box volume. The general thickness of the foam used for protection was about 1.5 inches per side. This is quite an achievement in a design of a minimal packing size for a delicate instrument.

11.1   Shipping Environments

The shipping environment includes all of the drop, shock, and vibration that occur with the transportation of the finished product to the customer. This includes warehousing and transport (air, ground, ship, cargo) to the customer.

There are standards to follow for transporting goods such as the IATA (see Ref. [1]). These include various issues of safety, maximum sizes, and weights.

So, just to further define *the item* that we are referring to in this chapter, the *packaging* can be referred to with the following terminology in this chapter:

The *packaging* is normally referred to as what the customer gets at the POS (or, "point of sale"). It includes the shipping box and packaging materials.

The *shipping box* can be a very complicated design in itself. It can involve custom molded foam, cut foam, die-cut cardboard, and graphics for the visible cardboard surface. The graphics usually can include:

- Part marking/bar coding
- Corporation logo(s) and corporation name
- Recycle symbols
- "Fragile" or "this end up" markings
- Marketing information

I'd like to distinguish the *shipping container* (from the *shipping box*). In this context, the shipping container would be a box specifically designed for vehicle transport. These would include wooden crates or other transportation boxes created due to unusually harsh shipping environments.

The *shipping box* can be one that the customer sees at the POS and thus will include graphics appropriate to that type of sale. Or the box can be very plain as it might be intended to be not necessarily seen by the customer at the purchase time.

First question in the shipping packaging design may be one of whether to use an existing design, a modification to an existing design, or a completely customized solution (much like the design of *any* part). The second question is to determine exactly what goes into the box which can include:

1. The "main electronic enclosure" assembly
2. External cables
3. Power supply or power supply adapter
4. Manuals, diskettes, and assembly instructions
5. Safety warnings

Another item to consider is the orientation of the product to the way the packaging opens. The items within the packaging should present themselves to the customer in a logical fashion.

11.2   Product Fragility

Fragility is a measure of the largest possible deceleration a product can withstand without breaking. "Without breaking" is a key phrase in the previous sentence. This can be defined by:

- Continues functioning (as it did before drop)
- Has no "visible" damage

Fragility is measured by various techniques including impact testing (ASTM D-3331) and shock machine testing (ASTM D-3332). A product's fragility determines the thickness of shock absorbing material required to protect it from a given impact load. Although a perfect shock absorber does not exist, it provides a useful theoretical yardstick. When a product is dropped from a height, H, a "perfect shock absorber" of thickness T returns a peak deceleration of:

$$H/T\,g$$

If fragility F is known and drop height is specified, the minimum thickness (of a "perfect material" would be:

$T_{min} = H/F$ (where T_{actual} will usually be between $2T_{min}$ and $6T_{min}$) per Ref. [2].

Shock energy is usually easy to determine. Where anticipated shock results from a free fall onto a rigid surface, shock magnitude is estimated from product weight, W, and drop height, H. This shock energy is then stated as equal to $W \times H$, which is the potential energy of the system (that gets converted to kinetic energy, $1/2\,mV^2$, at the drop).

As a *guideline only*, fragility levels could be estimated by (never substitute for test information) (see Ref. [3]):

- Extremely fragile: special military applications, precision aligned instruments, 15 – 25G
- Very Delicate: mechanically shock-mounted instruments and electronic equipment (shock mounts provided for in-service protection) firmly secured prior to packaging, 25 – 40G
- Delicate: aircraft accessories, computer equipment, and other office equipment, 40 – 60G
- Moderately delicate: video equipment, computer monitors, 60 – 85G
- Moderately rugged: equipment with minimal electronics, 85 – 115G
- Rugged: machinery, 115G and up

11.3   Material Selection

Shock absorbing materials are chosen by matching material firmness to impact severity. The idea here is simple: small shocks are best absorbed by firmer materials, while large shocks are best absorbed by firmer materials. Material firmness for a given shock magnitude can be selected by referring to impact test data or design curves provided by material suppliers. The two types of design curves used most often are:

Dynamic cushioning curve: summarizes impact test data for a particular drop height and sample thickness (peak deceleration, g vs. static stress, psi). Static stress is drop weight divided by impact area or W/A. The advantage of this approach is that the vertical axis gives peak deceleration directly. The disadvantage is that each curve applies to only a particular drop height and material thickness. For example, a curve can be shown for:

$$T = 1\,\text{inch} \text{ and } H = 12\,\text{inch}$$

Showing static stress minimized in the 50–75 g levels.

As H increases, say to 24 inches, static stress is minimized in the 75–100 g levels.

Cushioning efficiency curve (or *J*-curve): provides more useful information than dynamic cushioning curves. Cushioning efficiency curves plots J vs. U, where U is the ratio of impact kinetic energy to material volume absorbing the impact. U is also called the impact energy density and equals WH/AT. Peak deceleration estimates can be read directly for these curves for any set of test conditions, with regard to thickness or drop height.

In this method, peak deceleration data are plotted in curves of J vs. U.

J is the dimensionless ratio of actual to perfect peak deceleration; the closer J is to 1, the better the material performance. Each J-curve has a minimum value for a particular energy density, indicating the energy that the material can absorb most effectively. A minimum possible J-value depends on the type of material. Some examples are:

- Rigid foam, $J < 2$, used for single impact only as it is crushable
- Flexible foam and microcellular elastomers, J between 2 and 4
- Solid elastomers, $J > 5$

Let's put the above background to work to solve a typical problem: a portable computer must be designed to withstand a 6 inch drop. Its internal mechanical design is complete and only the shipping box remains unsized. We must select an appropriate cushioning material and box dimensions to fit the cushioning material. We are given a foam manufacturer's curves for their three foam types and densities:

Material	Density (lb/ft^3)	J (efficiency)	U (in-lb/in^3)
Flexible foam	2	3	1
Microcellular Elastomer	20	3	25
Solid elastomer	70	5	100

1. Determine fragility: to design the shipping container, a fragility value must be measured (or assumed). Without this information, two hazards are possible: underdesign and drop test failure or overdesign with accompanying cost and weight penalties.

   The test lab measures a fragility of 100G for the computer. That is, with a shock table set at 100G, the hard drive in the computer doesn't function properly (after the test). The computer functioned properly with the shock table set at under 100G. If a perfect material were available, the computer could be protected from a 6 inch drop by a minimum thickness of $H/F = 6/100 = 0.060$ inch.

If the computer were more fragile, say 20G, then 6/20 or 0.30 inch of perfect material would be needed.

2. Estimate shock energy: when the 10 pound portable computer is dropped for a height of 6 inches, its kinetic energy upon impact is $WH = 10 \times 6 = 60$ in-lb.

3. Select a material: materials with the lowest J-values for the energy density are candidates. Energy density, $U = WH/AT = 60/AT$. Because thickness is determined primarily by fragility, a perfect material ($J = 1$) requires Tmin = H/F, or 0.060 inch. If designing for a single impact, a crushable material could be used ($J < 2$) that has a thickness less than $2T$min, or about 0.12 inch. But the computer must absorb many impacts; therefore, a material is suggested that keeps $J < 4$. Thickness becomes $4T$min, or 0.24 inch.

4. Calculate energy density: the impact area depends on how the computer falls and how the shock absorbing material is supported within the case. Impact area of 5 in^2 results in $U = WH/AT = 60/AT = 60/(5 \times 0.24) = 50$ in-lb/in^3.

Impact area of 10 in^2 results in $U = WH/AT = 60/AT = 60/(10 \times 0.24) = 25$ in-lb/in^3. These values, 50 and 25 in-lb/in^3, represent the upper and lower limits on U.

Select materials whose cushioning efficiency curves satisfy $J < 4$ for U between 25 and 50. From the Table of Materials (above), flexible foam is too soft, and the solid elastomer is too firm. The microcellular elastomer has a J of about 3 for U at about 25. Therefore, if the microcellular elastomer is used in the computer case, peak deceleration is estimated with JH/T or $(3 \times 6)/0.24 = 75$G.

75G is less than the 100G fragility value, indicating an appropriate material, thickness, and area.

Another consideration of material selection for transportation for bags, cushioning, etc. is the material's ability to dissipate static electricity. Various standards exist for determining surface resistivity, volume resistivity, and decay.

11.4   Testing for Transportation

There are test specifications already existing for most transportation environments. Two examples of common tests are:

Loose cargo bounce test: this test is per National Safe Transit Association (NSTA) Project 1A, Section 1: Vibration test (packaged products under 100 lbs), Section B. complies with ASTM D-999 (see Ref. [4]). This test is used to simulate transportation as loose cargo over long distances and to aid in inducing and/or identifying failure modes in packaging. This is usually followed by package drop tests. Minimum of 14,200 cycles of sinusoidal vibration input with 1-inch peak-peak displacement. Frequency adjusted so that a minimum clearance of 0.062 inch is attained between package surface and shaker table for each cycle (usually <10 Hz). Package *is not* attached to shaker. A single 90° (or 180°) rotation to be done after ½ of the vibration period has passed. Vibration and inspection to be performed before drop testing. This testing should be repeated whenever packaging is changed or shipping-related field failures (DOA's) occur.

Free fall drop test: this is per NSTA Package Drop Specification Project 1A (packages under 100 lbs.). Section 2, drop test. This test is to be performed on a suitable drop test machine to insure reliable and repeatable drops. Define box surfaces as per specification. Perform drops in order defined in specification. Drop heights,

- 1 through 20.99 lbs = 30 inches.
- 21 through 40.99 lbs = 24 inches.
- 41 through 60.99 lbs = 18 inches.
- 61 through 100 lbs = 12 inches.

Define one (glued preferably) end corner as the 2-3-5 corner.

- Top = face #1
- Right side = face #2
- Bottom = face #3
- Left side = face #4
- Near end = face #5
- Far end = face #6

Ten drops. Drop order,

1. The 2-3-5 corner
2. Shortest edge radiating from that corner
3. The next longest edge radiating from that corner
4. The longest edge radiating from that corner
5. Flat on one of the smallest faces
6. Flat on the opposite small face
7. Flat on one of the medium-sized faces
8. Flat on the opposite medium face
9. Flat on one of the largest faces
10. Flat on the opposite large face

Inspect the package as needed throughout the test. Response measurements will be taken for the flat surface drops with the accelerometer mounted in a sensitive location along the axis of excitation within the packaged product. Appropriate low-pass filter may be used as appropriate. Note: the input to the packaged item will not exceed acceleration levels of those defined by the survival level shock specification for the product. This drop test should be preceded by the loose cargo bounce test as pre-conditioning.

Other testing besides flat and corner drops includes:

- Edge drop
- Edgewise rotational drop
- Cornerwise rotational drop
- Inclined impact
- Pendulum impact
- Tip-over test

- Rollover test
- Rolling impact test

References

1. Standards, manuals, and guidelines of the International Air Transport Association (IATA)
2. Arimond J, Shock control – better ways to select materials. Rogers Corp, Article from Machine Design
3. Gemini cases catalog, 2001
4. Vibration testing of shipping containers, ASTM D 999-75

Chapter 12
Documentation

This chapter will review the essential need for documentation and the various types of documentation.

All of our ideas need to take a "written" form for them to:

1. Be understood and (perhaps) followed in a repeatable manner.
2. Serve as a basis for continuous improvement of those ideas.

To that end, it is reasonable that our enclosure designs, which start out as ideas, be turned into this "written" form. In the current times, this written form is really one of a digital (electronic) form. We use computers to create, file, analyze, and generally document our work. We've become creative in our ability to:

- Use pictures and video to augment our ideas.
- Use search engine information from the Internet to augment our ideas. This now includes information from the entire (Internet accessible) world.

We'll start the chapter with a definition of what the general traits of all of our documentation should be. We'll explore all of the types of documentation including assembly drawings and parts lists. We'll see the need for good "version control" of our documentation, that is, as the documentation changes; we need a solid system to make sure everyone knows what the latest version is, and just what that latest version is.

The need for solid documentation of our project, leads to a discussion about how that documentation is shared among the various groups who have a "need to know" in our project, and I title this Sect. 12.5, "Linkage Between Groups."

We'll end with a discussion that shows how a "normal" process of a project goes from "engineering control" to "manufacturing control." Thus, the project, in the form of the "documentation of the project," gets transferred in an orderly fashion from design to manufacturing.

© Springer International Publishing AG, part of Springer Nature 2019
T. Serksnis, *Designing Electronic Product Enclosures*,
https://doi.org/10.1007/978-3-319-69395-8_12

12.1   EPE Designer Output

The EPE Designer has made several contributions to the overall product delivery process. They have come up with many of the salient ideas that are keys to the design's success. They have prototyped and tested those designs.

What the EPE Designer will leave behind, as they move on to the next project, will be all of the documentation for this project. Let's talk about this legacy documentation.

Documentation should be:

1. Unambiguous – very clear and not open to interpretation
2. Follow "industry standards" for content – helps for uniformity and clearer universal understanding
3. Concise – leads to increased productivity
4. Easily transferrable into "standard" formats and language of the receiver of the information
5. Easily updated with new, additional, or corrective information – change control should be transparent and easily followed

There are various standards available for documentation procedures including:

- *Drawing Requirements Manual*, published by Bishop Graphics, Inc. In that manual is a listing of specifications and standards data, which include references to:

  - Department of Defense (DOD) specifications (DOD-D-1000)
  - Military standards (such as MIL-STD-454)
  - American National Standards Institute (ANSI) specifications (such as ANSI Y14.1, already referenced)
  - International Standards Organization (ISO) standards such as ISO 1000

Also, there are various *company standards* that are widely being followed. It is clear that any business entity needs to provide written procedures, standards, and guidelines for their own documentation processes they themselves will be using.

The designer of enclosures will be documenting a variety of items; some of which have already been covered in this text:

1. Assemblies and the parts lists contained in those assemblies (detailed in Sect. 12.2).
2. Individual parts – this is all of the information about fabricating each part. It includes:

   - Geometry of the part
   - Material and finish of the part
   - Inspection criteria of the part

3. Testing information (see Chap. 6 on Assembly Testing and the environmental testing requirements of Chap. 7).
4. Changes to documentation (see Sect. 12.4, Version Control).

5. Supplier, cost, and lead-time information (see Sect. 12.5, Linkage Between Groups).
6. Other "project management" documentation (referenced throughout this text).

12.2 Assemblies/Parts Lists

Some would argue that the "assembly" is the most critical and most creative element that a designer invents. Of course, with larger assemblies, it is possible that a single part (of that assembly) is actually the most critical. In any event, a designer must be thinking of the overall assembly even when inventing a single part. The documentation of an assembly consists of:

1. The assembly itself, in pictorial form – the assembly drawing can show the assembly as:

   - Final position, as in the assemblies intended use.
   - "Exploded view" or shown, taken apart. This method helps the reader visualize the general steps of assembly.

2. A parts list. The parts list may be actually a physical list on the assembly drawing or its own separate piece of documentation. Usually, there are references coordinating the separate parts shown on the assembly, with the listing of the separate parts. For example, an attachment with an arrow pointing to "Part 2" of the actual assembly will be associated with "Part 2" on the parts list.

   In some cases, there is also a document that explains (with words and pictures) the actual assembly process. This document (perhaps referred to as the "assembly process") is coordinated with the assembly drawing and the parts list.

   A parts list can also be referred to as a bill of material, or simply a BOM.

12.3 Part Numbering

Each company devises a part number scheme to suit their own needs and processes. They are in one of three types:

1. Intelligent – the numbers have an assigned "meaning." For example, all parts that start with a "9" are "fasteners." Or, all assemblies on the "Orion Project" are marked with the extension "Orion" (as XXXX-Orion).
2. Consecutive (or non-intelligent) – part numbers are consecutive to each other and have no particular association with any other part.
3. Some blending of the above two systems.

So, *just as an illustration of part numbering* and going back to "assembly draw-ings" in Sect. 1, we might have an assembly (12345-120, which is the 120 V version of a computer) that consists of:

1. 12345-XXX-AD – the assembly drawing used for both the "-120" and "-220" versions
2. 12345-120-PL – the parts list for the "-120" version
3. 12345-XXX-AP – the assembly procedure used for both versions
4. 224-120 – the individual 120 V power supply which would be called out on the 12345-120-PL parts list

12.4   Version Control (or Revision Control)

"Nothing is sure as change." That statement is certainly true of documentation. Companies must possess change control systems that have the following characteristics.

12.4.1   Change Visibility

Reviewers must be able to see the difference between the present revision level of the documentation and the proposed revision level. The initial revision level of doc-umentation is nominally stated as "Revision 1." So, their "part numbers" (see Sect. 11.2) actually are not complete without the revision level. Thus, in the example of Sect. 12.3, the parts list would actually be:

12345-120-PL Revision 1

The next change (unapproved as yet) would take this to:

12345-120-PL Revision 2

So, anyone who is looking at the documentation for this parts list would see that the present revision level is "1" and that there is a "pending change" that would take the revision level to "2." The viewer would see both revision levels and see the "unsigned-off" status of Revision 2.

It is very common to delineate between "unreleased" (or "preproduction" parts) and "released" (or "production" parts). This can be done by using a "number" for revision level (as in above example, 12345-120-PL Revision 2), where the notation Revision 2 denotes an "unreleased" assembly. As the assembly matures and gets to the "released" stage of its product life cycle, it would change revision to Revision A and be known as 12345-120-PL Revision A.

In that manner, it is clear to just about everyone that the assembly 12345-120-PL is either "unreleased" or at a "released" status. A "released" status usually has dif-ferent levels of review.

12.4.2   Parts Disposition

All parts need to be clearly identified as to their revision level and if they can be used "interchangeably" with the previous revision levels. Sometimes, to avoid issues that arise with this "interchangeability," a part must actually change its "entire" part number to avoid the wrong part being used. At some point, a "disposition" must be determined for the old revision level (Revision 1). It must be determined how to use the present stock on hand (and the stock "in transit") of Revision 1. Some choices are:

- Use stock on hand (as is).
- Rework stock on hand (with the specified rework).
- Scrap stock on hand.
- Not affected.

An evaluation of the impact of the change to the entire build process must be made. This includes:

- Cost of the change
- Form/fit/function (backward/forward compatibility issues)
- Customer specifications
- Production schedule
- Where used (exactly where the part or assembly is used or called out)
- Service implications (product recall, field upgrades, build-out of old/obsolete assemblies)

12.4.3   Nimbleness

As the sheer number of changes that can occur as the design gets prototyped, tested, and released to production, it becomes critical to process the changes very expediently. Some of the parts have very long lead times needed for their ordering, production, and shipment to assembly; therefore, change control needs to occur rapidly. For example, a change creating 12345-120-PL Revision 3 (see above example) can complicate the process even further if Revision 2 hasn't had complete sign-off as yet.

12.4.4   Token Ring

Similar to the network topology, the information must be able to be passed along to various corporate groups (see Sect. 12.5) for review and comment. Some of the people reviewing the changes are actual (signature) approvers of those changes, while others are receiving the information for reference purposes.

12.4.5   Adaptability

Change control needs are different in the different phases of the product life cycle. As the project matures, more review of the changes is required. In the early stages of prototyping, speed is more important to push toward larger build needs. As the quantities of parts ordered gets larger, the number of parts "in process" gets larger, and the need for review and supplier intervention increases.

12.4.6   Implementation

Documentation needs to be updated and distributed upon approval.

12.5   Linkage Between Groups

An excellent product is the result of various functions working together to define the specifications of individual parts and their subsequent assembly. Some larger companies have a "new product introduction" group, while smaller companies may not have a group of that charter in their formal process. So, groups will need to work together to advance projects toward a release to manufacturing (Sect. 12.6). A "linkage" is required between the following groups that absolutely DEPEND on *crisp* documentation and *stellar* change control processes:

- New product introduction (if this group exists)
- Engineering
- Purchasing
- Project management
- Manufacturing test
- Manufacturing pilot build (if separate from production build)
- Manufacturing production build
- Quality control
- Service
- Marketing
- Sales
- The outside parts fabricators (yes, the suppliers must be a "partner" in the overall development of individual parts)

We have seen this listing of groups involved in "the project" a few times in this text, and right now I am emphasizing the importance of *documentation* that must exist in place that each of these groups can access. This is normally done in a large database material resourse planning (MRP) and demand forecasting enterprise solutions. The technology and software are changing extremely quickly in this area. "SMAC stack" is an example (see [1]):

"A manufacturing comeback is being driven by SMAC – social, mobile, analytics and cloud. The SMAC Stack is becoming an essential technology tool kit for enterprises and represents the next wave for driving higher customer engagement and growth opportunities. The need to innovate is forcing cultural change within a historically conservative "if not broke don't fix it" industry, and SMAC is helping early adopters in the manufacturing market increase efficiencies and change."

12.5.1   Engineering Deliverables and Needs

Obviously the above groups work together with stated *needs* and their "*deliverables.*" Main *deliverables from Engineering* are:

1. Documentation of the product. Includes individual parts, assembly information, and parts list.
2. Test requirements for the product (for final customer shipment).
3. A product that meets or exceeds the product specification and a test report that states so.
4. Viable (and agreed to) fabrication techniques for all of the individual parts.
5. Viable (and agreed to) assembly techniques for the subassemblies and above. This includes all "assembly fixtures" needed to assemble the product.

To be able to deliver the above "deliverables," the *needs of Engineering* are:

1. A representative from purchasing responsible for new product introduction on each project. That person(s) is the point of contact on all purchasing matters. Going forward in this section, the "new product introduction team" is the team within the organization that has the responsibility to bring *this* new product to the marketplace. In this text, we will refer to the team as new product introduction or NPI. The membership of NPI varies from company to company but can be comprised of members from engineering, purchasing, and manufacturing.
2. Engineering purchases (purchases made before production release):

   On every purchase requisition, the project needs to know:
   Supplier – this usually comes from either:

   A. Approved vendor list (AVL) – production vendors already approved for usage
   B. Prototype shops – vendors who will do short runs with quick turnover
   C. New vendor – is "developed" that has strong potential to be a part of the AVL

The selected supplier should make sense for the company as the vendor should be the best combination of quality/cost/responsiveness/technology. It takes a lot of time and effort to survey the vendors and approve them for production. Often, a competitive bidding process between vendors is used to make the selection.

Some vendors are better suited toward "production" (high quantities of parts over a long time period commitment), while other vendors specialize in short-term, quick-to-deliver parts that may be best suited for a project prototype phase.

Date material needed – this is an agreement between what is needed by the project and what is realizable from the vendor. Deviations from expected delivery date must be known. Monitoring should be done to see if cost or time improvements can be made. It needs to be established as to who is the point of contact between vendor and this company.

Tooling requirements – this is a critical purchase in that it usually involves high costs and long lead times. NPI is responsible for the competitive quoting process, payment schedules, legal recording of the tooling, and progress reports. Tooling changes (a change to the part design that causes a change to the tool) must be carefully evaluated for the effect on schedule and costs.

Costing information – costs change for an individual part based on:

- Quantity purchased – each company needs to determine how "quantity" dictates pricing. They must generally choose between (quantities shown are just "a typical example"):

  1. Initial "prototype" run of parts (10 @ $100 each)
  2. Pilot run of parts (100 @ $20 each)
  3. Initial "production" run of parts (1000 @ $5 each)
  4. Initial "steady state, spaced delivery, contracted" (12,000 parts committed to but delivered at 2000/month for 6 months, @ $1 each)

So, obviously, the pricing goes down as we go from 1 to 4 (above), and each organization chooses how their MRP system will actually list the price for the particular part. So, the MRP system would probably never have the $100 pricing listed as this is unrealistic but have a more realistic pricing such as the $5 price for this example part and evolve to the $1 price at a later date.

- Tooled vs. un-tooled parts – part cost reduced as tooling is utilized for fabrication. (See previous discussion on ROI.)
- Vendor selected – each part will need a "request for quote (RFQ)" where NPI will choose best vendor based on metrics determined (price/quality/delivery).
- Shipping costs – fairly standardized practices but needs to be considered.
- Expediting fees – usually not considered with standardized costing.
- Design changes – easily tracked per revision change to the part.
- Supply chain changes – this includes modifying the amount of work that each supplier "value-adds" to the part. As an example, if we modify a panel to include the addition of a label to that part (to be installed by the fabricator of the panel), we are removing the purchasing of the label and the assembly of the label from "in-house" to the panel fabricator. So, the panel may go from $1.00 to $1.10 as it is now an "assembly."

All of the above factors are influencing how much a part costs at any particular time, and the project must know what the cost information is at various snapshots during the project. For example, if production quantities are lowered, the cost per part will increase. The NPI must be continually evaluating potential cost savings ideas during development and beyond.

12.5.2   Long-Term Production Purchases

NPI needs to coordinate the purchase of parts needed for pilot/prototype builds *and* production builds. At some point in the product design process, there is a "handoff" between engineering and manufacturing (see Sect. 6), and that handoff needs to be as coordinated as smoothly as possible. This point is a transition from "engineering-driven" purchases to "production-driven" purchases. This may include using production fabricators as early as possible (as opposed to prototype shops). NPI needs to manage material demand beyond the planned product introduction date. This requires knowledge of material lead times and build plans. NPI needs to be very well integrated with the project team as new parts are added to the BOMs (even with incomplete part information) which could lead to future material shortages. The *NPI responsibilities* include:

1. New parts information: The NPI is responsible for getting all part information into the material recourse planning (MRP) tool to suit the project's needs. Engineering to submit original "new parts form" with initially known (engineering) information. The new part is added to the BOM.
2. Updated BOM information: The NPI is responsible for getting the parts lists, assemblies, and subassemblies into the MRP system. This would include Revision 1 originally submitted by engineering and all subsequent updates. All subassemblies are identified, and the material supply chain for each part and each subassembly is documented.
3. Pilot builds: The NPI is responsible for "kiting" of parts required for pilot/prototype builds. By "kiting," I am referring to the sequestering of parts needed for the build. If ten assemblies are to be built, then ten sets of parts need to be put aside and "earmarked" for the build. For the build, NPI will make sure the parts are ordered *in time* to allow the build to proceed on schedule, report shortages as appropriate, and insure that parts meet specifications for those parts.
4. Assembly processes: NPI to assure that a build team is scheduled for assemblies and that they have:

   - Build documentation
   - Build assembly fixtures
   - Test procedures and equipment

12.6   Engineering Release to Manufacturing

Each company must determine the best process and procedure to move a product from what is considered "in development" to a product that would be considered "in production." I'm generally titling this transition as the engineering release to manufacturing.

The following can be considered a checklist of items that are required from engineering to release a product to manufacturing.

Five categories of *requirements* are defined:

1. Deliverables – these are documents that will reside in the corporate MRP system. All documentation is complete and ready to release. This documentation has been reviewed by manufacturing and quality assurance. Revision level will change from "engineering revisions" to "released revisions" as noted previously, at Revision A. This documentation includes:

   - PCBA – printed circuit board assemblies and their individual parts. Includes the parts lists (BOMs).
   - Product assemblies – their assembly drawings and their individual parts. Includes the parts lists (BOMs).
   - Purchased parts and assemblies.
   - Test specifications (both for PCBA and box build).
   - Engineering to participate in the production of both the user's manual and the service manual. Service parts are identified and documented.
   - (Software and firmware is not being addressed by this design text.)
   - All parts and assemblies have all information required for purchasing, stocking, servicing, and inspecting of those parts and assemblies.

2. Design verification – this is a documentation that records the passing of (or exception to) testing performed by engineering to meet the product requirements specification. This includes environmental testing done to the individual parts and assemblies.
3. Agency approvals attained – this includes all documentation and certification from both laboratories and testing facilities. All labeling is on the product and its packaging in the approved manner. Documentation resides in the corporate MRP system.
4. Long lead items on order – engineering takes responsibility for the identification of parts needed for future (scheduled) builds. Engineering will work with manufacturing to ensure that parts with lead times identified that place them in critical path to enable future builds will be ordered in time for those future builds.
5. Tooling – ownership and responsibility for tooling is transferred to manufacturing. This would include tooling developed for injected moldings and castings. Other assembly tooling such as special drivers or assembly tools which might be codeveloped with manufacturing would become the responsibility of manufacturing at this time. Assembly fixtures will become the responsibility of manufacturing. Test fixtures (used in assembly testing) will become the responsibility of manufacturing.

Sign-off will occur when representative members of engineering and manufacturing agree that the above requirements have been met.

Reference

1. Kotelec M., 5 Manufacturing trends that will shape the market in 2015. Verizon Enterprise Solutions.

Chapter 13
Continuous Improvement

We will finish the text with thoughts toward the need to remember both the things that worked well and the things that didn't work so well. This should occur both in the design itself and in the design process. In fact, we will not only remember them, we'll write down those memories. Every endeavor worth doing, and the EPE Designer certainly works on those endeavors, is worth analyzing on how to improve. The thought here is that, before tackling that next project, we'll "fold in" what we learned on this latest project. The entire project team should analyze in a formal manner the steps needed to make the next project better.

This chapter is organized by starting with a discussion on the need for metrics – we can't say we need to do something "better" until we have a metric to compare against. We'll revisit the design review practices that were done during the project. Many companies have formalized continuous improvement, and we will look at some of them that are out in the literature. We'll look at a checklist that can help us improve a process (in this case, the general process of cable design). Finally, we'll look at the need to create and iterate the general procedures that will be used by the company.

13.1   Improvement in Design

The word "improvement" implies the need for metrics. If we say that we would like to improve, we don't know if we *have* improved, unless we compare what we are attempting to improve, with an existing baseline.

Improvements may be difficult to quantify. Some of the basic items in the product design world that one would think of to improve would be:

- Less cost
- Less time needed from concept to customer ship
- Less product returns or warranty issues

© Springer International Publishing AG, part of Springer Nature 2019
T. Serksnis, *Designing Electronic Product Enclosures*,
https://doi.org/10.1007/978-3-319-69395-8_13

An item such as improving "less cost" can even be quite involved. One would have to find the cost billed from the vendor (at a particular quantity and delivery schedule) and include "hidden costs" such as shipping costs (where delivered) and costs related to quality control. It would only be after knowing the "present cost" could one compare the "new, proposed cost."

We could also look at improvements in our designs such as:

- More creativity
- More satisfied customers
- Increasing the productivity seen by our customers from our product

It would certainly be difficult to quantitatively measure "increased creativity" in a product, but customer satisfaction can be measured. As to making a product "more productive" to the customer, one would have to know exactly how our customers use the product (which can be measured).

In this Sect. 13.1, *Improvement in Design*, I want to talk about the ability to improve the design of an *individual part* or an entire product. While in the Sect. 13.4, *Improvement in Processes*, I want to talk about the ability to improve a *process* in design. For example, if a process is in place to accomplish a company goal, that process itself can be improved.

13.2 Design Review Practices

We have previously talked about the need for design reviews in Chap. 1. It is hoped that pointing out the value attained from them, at both the beginning of the text and at the end, will leave the reader with lasting memories. Analyzing the design review process for how well it worked during the project, and what the team needs to do to make them better, should be a part of continual improvement.

Design reviews are one of the most positive steps that can be taken to improve an idea or design. They can be "formal" or "informal," a preplanned step in the product design process, or a "spontaneous" meeting at your workstation. The process of having another person review your work will improve your work. For one thing, you are usually forced to talk about your design and how it meets certain goals. Any design review that results in a better design is good, but, there are "best practices" for design reviews that seem to maximize gains. Some of which are:

1. Deciding when to have the design review. Have them often (as often as they continue to have value). They can be formal or informal. Informal reviews can happen whenever ideas are needed. Formal reviews are likely mandated by a milestone in the product design process.
2. Preparation required? A "formal" review needs a lot of preparation – you must be able to explain (in whatever detail as needed) just about anything anyone asks about the design. You have to be willing to "put work into" the preparation so that you will get "maximum value out." You can certainly "bound" (place limits on) the design review to confine the conversation to a subset of the entire prob-

lem. Preparation should be such that it is demonstrated that the design satisfies the design requirements and was chosen by you quantitatively from among a few viable alternatives. One of the goals of the design review would be to have the actual design *improve because* of the design review. That is, if the design is improved, or its present state is affirmed that the solution meets the product requirements, it has been a worthwhile design review.

3. Deciding who should attend. Two schools of thought exist on this. Having a peer review from those that are very familiar with your area of design is good. These people will be very familiar with both the problems you are solving and the tools available to solve those problems. The more direct experience that they have, the better they will be able to relate those previous solutions to the design being reviewed. However, it also seems to be of value to have people who may be technical or are *not* in the direct field of expertise of the design. Those people will come at some solutions "from another area," that is, they come at solutions "from another direction." If "round seals" are normally used, they are the person who may suggest "...why not a square seal?" So, the design review team can be cross functional and can include people from outside of the "normal" team.

4. Deciding how many people to attend. There does seem to be a "right size" to a design review. Too few people can result in not enough ideas (or honest criticism) being expressed. Too many participants can lead to loss of focus from everyone in a large group as everyone can't "get engaged" with the issues at hand.

5. Meeting etiquette. Make sure that everyone knows that all ideas are welcome and divergent ideas aren't all bad. Active participation should be valued. All of the discussions should be directed toward everyone in the room (avoid "splinter group" conversations).

6. Meeting logistics. There must be a prepublished agenda. Meeting notes should be recorded, published, and distributed. Items noted that need follow-up, should be followed up. There should be a known length of meeting. Time is of the essence – do not deviate from the meeting objectives and it may not be necessary to (completely) solve a particular problem at this design review. There must be a **leader** of the design review, and that person must keep control of the meeting to stay focused on the agenda and note important items brought up.

13.3 Existing Improvement Methodologies

Several "industry standards" already exist along the lines of this general subject of continuous improvement. One is Six Sigma (from [1]).

The textbook definition of Six Sigma refers to a step-by-step approach designed to reduce the number of errors in a given process. At the Caterpillar Corporation, anything that produces a result is considered a process. This can mean everything from assembling a product to finding areas of opportunity to profitably grow the company. The end result is fewer errors and more successful, efficient processes across the enterprise.

The Greek letter *Sigma* is used to represent a statistical unit of measurement. It defines standard deviation or, in Caterpillar's case, the number of errors made in a product or process. Six Sigma equates to only 3.4 errors per million opportunities.

Most businesses operate at a Four Sigma level, meaning they're about 99.5 percent defect free. Sounds pretty good, right?

But it really isn't good enough when "99.5 percent defect free" translates into the following statistics:

- Two accidents a day at major airports
- 210 min a month without electricity
- 2500 incorrect surgical procedures a week
- 20,000 pieces of mail lost an hour
- 100,000 incorrect drug prescriptions a year

To reach the Six Sigma level, Caterpillar uses "Six Sigma methodologies" to guide projects and implement processes. The methodology utilizes a common, disciplined approach that focuses on gathering information, analyzing data, and making fact-based decisions to ensure the most efficient and effective processes, products, and services are in place to meet customer requirements.

Other methodologies are:

- Design for Six Sigma (DFSS)
- Lean Six Sigma (LSS)
- DMAIC (Define – Measure – Analyze – Improve – Control).

All of which provide systems to reduce errors and gain productivity.

13.4 Improvement in Processes

Continuous improvement can be made to a *particular* single part, a subassembly, or assembly. Also, a continuous improvement can be made into a generic process or a *generic* "class" of parts, subassemblies, or assemblies. For example, it can be decided to reduce the cost of (all) cables in a particular design or even all cables being bought by a group or the entire corporation. One would first decide the "scope" of the effort. Going forward with our example of reducing cost in purchased cables, say for the corporation, one would:

- Design a methodology to understand the cost components of a cable. This would include all of the cost components including parts, assembly, testing, shipping, overhead, and profit (margin) of the cable manufacturer.
- Look at quantity purchases of components or assemblies.
- Look at revising the suppliers involved with cables. This could be reducing suppliers to gain quantity pricing or even increasing suppliers to get the appropriate cables into different (more appropriate) cable vendors.
- Look at the potential cost reduction involved (of this effort) and compare those savings with the cost of the cost reduction effort.

- Look at cable commonality such as testing, assembly, and quality control efforts. Can cables be produced at a lower cost if we create standards for the testing, assembly, and quality control items?

As an example of a process, in this case, cable design and documentation, the following is shown (abbreviated version to show overall checklist considerations):

1. Purpose

This document is intended to provide guidance for the design and documentation of cables. The design must fulfill the requirements of the functionality intent of the cable, and the methodology provided by this document should aid the design process.

The cable documentation should provide a consistent and clear expression of that design intent, while meeting company drawing standards. Cable documentation should "stand on its own" and be free of "tribal knowledge" allowing the cable to be quoted at any company-approved cable vendor.

A design checklist (Sect. 3.4, below) is in this form of questions that a cable designer should review so that common design considerations are met.

2. Scope

This checklist can be used for any company cable. However, certain documentation processes vary from company site-to-site.

3. Design Process

   3.1 Cable components: Cables generally consist of:

   - Connectors
   - Wire (raw cable)
   - Strain relief
   - Dust caps with lanyards
   - Shrink tubing
   - Labeling

They can also include many custom parts including built-in PCBAs, custom molded parts, and custom sheet metal parts. This checklist is intended to cover the most general of cables with a limited number of individual BOM components.

   3.2 Design responsibility: Cables are designed as other components of systems, in that they are reviewed for their engineering, operations, quality, and marketing considerations. They somewhat differ in that cables have both electrical and mechanical design characteristics. However, they are usually documented by the mechanical engineering function.

   3.3 Design and development process: Cables are individual assemblies that follow the normal product design and release process in that they are:

   - Conceived as a project need
   - Specifications for them are formalized

- Prototyped
- Tested
- Released to production
- Monitored for quality and cost

3.4 Design checklist (designer to check each consideration as complete and/or add additional comments):

1. Can the entire new requirements for a cable be attained by a present company cable? Has a search been done for similar functionality?

3.5 Standard notes:

1. Test requirements need to be exactly stated on the documentation. There can be test requirements for: Continuity (shorts and opens) – this test is normally required for every cable (100% tested).

3.6 RoHS compliant cables:

1. If cable is to be RoHS compliant, there should be a note that states "Completed Assembly to be in Compliance with the Reduction of Hazardous Substances (RoHS) per European 2002/95/EC."

4. Documentation Requirements

- 4.1 Drawing title
- 4.2 Assembly part number
- 4.3 Picture of cable
- 4.4 Wiring diagram
- 4.5 Drawing BOM
- 4.6 BOM in material resource planning (MRP)
- 4.7 Table of part number variations
- 4.8 General information regarding notes

5. Reference Documents

13.5 Guidelines Versus Procedures

All processes can be streamlined and made better. Tremendous cost reductions can result in improving any process at a company. A product life cycle process (any process that tracks products thru design, manufacturing, and end of life) is an example of a process that every company has in place. An improvement in this type of process helps *every* product being designed so that this type of improvement is perpetual.

So, if there isn't a company-wide (written) procedure presently in place for a particular repetitive function, that procedure needs to be created. It might be best to at least get that procedure in place and revise it as more information and experience is attained. Therefore, a two-step procedure is needed:

1. Step one – Create the procedure. This includes statements about the ownership of the procedure and how the procedure can be modified. Get input for the procedure from all groups that will be affected by the procedure.
2. Step two – Revise the procedure to make it be more efficient or better suited to help more related problems. There should be a balance between simplicity and coverage of "corner cases." Each time the procedure is used, the procedure should be reviewed to see where improvement in the procedure could be attained.

Procedures should look for ways of listing:

- *What* needs to be done (the goal or task) and a step-by-step detailing of how one would expect to complete the goal
- *Who* needs to be doing the particular item
- A timeline as to *when* individual milestones generally lead to overall goal completion
- *Reviews and approvals* needed to proceed to the next milestone

Finally, some differentiation about *approved procedures* vs. *guidelines*:

Guidelines allow participants to vary their process in what manner will best suit the particular problem or project. The general "tone" of a guideline is less formal than an approved procedure.

Approved procedures should be (must be) followed. Exceptions to the approved procedures would have to be noted as such. It's possible that entire sections of "approved procedures" would be labeled as "not appropriate," and the project could move on from there. The generic procedure itself would have very formal sign-off for any change to that procedure. The general "tone" of an approved procedure is "these procedures will result in excellence - do not deviate from them unless they are not appropriate for the task-at-hand".

Reference

1. McBroom H, Six Sigma: Foundation of Quality at Caterpillar. Peoria Magazine